Cover Description

On the left is a Bertini fresco of Galileo Galilei and the Doge of Venice, who was the elected chief magistrate of the Italian republic of Venice at that time. Galileo is showing him how to use the telescope.[1]

On the right is a rendition done by Arthur Avary depicting one impression of what the afterlife environment is like—a place of great beauty inhabited by spirits of the deceased.[2]

The galaxy shown across the top of the picture is an ultraviolet picture of the Andromeda Galaxy (also known as M31) taken by NASA's Swift mission.[3]

The pictorial metaphor I am trying to portray here is that our deceased loved ones are around us all the time, even though we can not sense them with our five senses nor detect them using the most sophisticated instruments developed to date by modern science.

Extending down from the cosmos and marking the separation between the realms of science and spirituality is the pendulum, used there to signify a point of meeting as well as a point of division, and which I also use as a metaphor of the oscillation between these two paradigms in our individual and collective experience.

---

[1] This image is a file taken from the Wikimedia commons website.
[2] Arthur Avary is a graphic artist who also does digital painting.
[3] I have included website addresses to the sites that describe this image and how it was obtained at the back of this book. It may be of interest to some that, in a universe where most galaxies are moving away from us, the Andromeda Galaxy is on a collision course with the Milky Way Galaxy.

THE GALILEAN
PENDULUM

A NEW SCIENCE REVEALS AN UNSEEN WORLD

BILL KASPARI

AuthorHouse™
1663 Liberty Drive
Bloomington, IN 47403
www.authorhouse.com
Phone: 1-800-839-8640

© 2013 by William John Kaspari. All rights reserved.

No part of this book may be reproduced, stored in a retrieval system, or transmitted by any means without the written permission of the author.

Published by AuthorHouse   03/20/2013

ISBN: 978-1-4817-0983-5 (sc)
ISBN: 978-1-4817-0980-4 (hc)
ISBN: 978-1-4817-0984-2 (e)

Library of Congress Control Number: 2013903887

This book is printed on acid-free paper.

Because of the dynamic nature of the Internet, any web addresses or links contained in this book may have changed since publication and may no longer be valid. The views expressed in this work are solely those of the author and do not necessarily reflect the views of the publisher, and the publisher hereby disclaims any responsibility for them.

Endorsements

Following the devastating loss of his 22-year-old son, Bill Kaspari describes his odyssey from unconcerned skeptic concerning an afterlife to firm believer. This is an inspiring work growing not out of conventional religion but the world's best paranormal research. Kaspari introduces the reader to many of the biggest names in the field and shows how his convictions gradually fell into place. Starting with nothing, he ends with a vision of life's ultimate purpose, which is far from finished when we die. A complete book, well written, and with glossary and index.

Stafford Betty, Professor of Religious Studies,
California State University, Bakersfield
Author of *The Afterlife Unveiled* and *The Imprisoned Splendor*

The Galilean Pendulum is the story of how an unseen world revealed itself to Bill Kaspari after the death of his second son. Kaspari invites readers to join him on his journey to understand life and death and find meaning in them. As an electrical engineer and former president of a medical electronics company, Kaspari approached this quest by turning to the world of science for answers. What he discovered gave him hope—and then confidence—that his sons did indeed survive their deaths, and are living in a world of which ours is but a dim shadow. The Galilean Pendulum provides a tour of the methods and findings of scientific research into the question of what happens to us after death, and places these findings in a larger context of theory of what it is to be a human being.

Bruce Greyson, M.D.
Carlson Professor of Psychiatry & Neurobehavioral Sciences
Director, Division of Perceptual Studies
University of Virginia Health System

In this thoughtful, analytical and well written book, the author, along with many other people in a similar position, had wondered if he would ever see, or contact his deceased son again and has tried to make 'sense' of his death.. He has taken pains to seek out evidence which would lead his thoughts along a new and inspiring path of hope through knowledge. Along with the author's own experiences and the vast amount of reliable data collected in many avenues of psychical research over the past 120 years, it seems to me (having researched these matters for nearly 30 years) that the evidence is more than sufficient, as in a court of law, to convince anyone, devoid of all bias, that the death of the body is not necessarily the death of that personality.

Tricia J Robertson
Lecturer,
DACE Tutor, in Psychical Research, University of Glasgow.
Immediate Past President of the SSPR

This book is dedicated to my sons John and Ricky and to my mom and dad.

Preface

This is the story of how an unseen world began to reveal itself to me in the years following the death of my son. It tells of the transformation I went through; how I was taken from viewing life through a lens which allowed me to see only the physical world to a vantage point where I was given a much broader view of our existence.

I invite the reader to take the journey I found myself on; what for me was a paradigm shift in my understanding of life. This journey began at a point where my feelings were a mixture of hopelessness, deep sorrow and frustration. I began a search trying to find answers to the question of what lies behind our existence, and in particular to try to determine whether or not I would ever be with my son again.

I am not a religious person, so I began to search for answers in a way that satisfied my scientific mentality and was surprised that, after a great deal of searching, both in the literature and in my own soul, the answers gradually began to appear. I discovered what for me is a new reality—a much better, far more interesting and more cheerful one than I ever had before.

Contents

Preface ix
Foreword xiii

Chapter 1: My View of Our Culture 1
Chapter 2: The Early Days 7
Chapter 3: Time For A Change 15
Chapter 4: The Near Death Experience 19
Chapter 5: A Glimpse of Other Paranormal
           Phenomena 25
Chapter 6: The Study of Mediumship 37
Chapter 7: Playing Charades 53
Chapter 8: Three Special Messages 71
Chapter 9: Understanding Mediumship 79
Chapter 10: Fortune Tellers, Frauds or Gifted? 89
Chapter 11: What conclusions can we draw from this? 95
Chapter 12: Memory 99
Chapter 13: A Scientific Look at Paranormal
            Phenomena 103
Chapter 14: Materialism versus Dualism 115
Chapter 15: Skepticism 123
Chapter 16: Science and Spirituality 137
Chapter 17: Letting Us Know They Are Around 157
Chapter 18: The Pendulum Begins to Swing Back 165

Glossary of Terms 173
Appendix A: Attribute Definitions and Scoring Values 187
Appendix B: Book Review 189

Recommended Reading 193
Bibliography 211
Website Addresses & Information 213
Acknowledgements 217
Index..223
About the Author 235

Foreword

The death of a child is one of the greatest tragedies loving parents can endure. Bill Kaspari, and his wife Diane, experienced this twice—first with their son, Ricky, when he was just an infant, and second with their son John, when he was twenty-two. The consequence of such losses is not additive pain; it is multiplicative. In the process of coping with John's unanticipated death, Bill was propelled to attempt to understand whether life continued beyond the grave. Given his background in engineering and leadership (including his having served as President of a Medical Electronics Company), Bill approached this great quest in a deep and methodical way.

I had the privilege to come to know Bill and Diane and assist them on their journey of discovery and healing. As revealed for the first time in this book, I arranged for Bill to have multiple readings with a few highly skilled and gifted psychic mediums who were participating in research in my laboratory at the time. These readings were novel in that Bill's deceased loved ones—including his father as well as his children—were "brought" to the readings via the late Susy Smith, a former journalist who became an authority on psychic phenomena and life after death. Susy became one of our "departed hypothesized co-investigators" whose research "from the other side" is detailed in my books *The Truth About Medium* (2005) and *The Sacred Promise* (2011).

As you will see, these controlled readings not only provided compelling evidence suggesting that the Kaspari's children's consciousness had survived physical death, but that they were continuing to grow and evolve. Moreover, Bill's systematic approach to the analysis, scoring, and interpretation of these readings contributed to the advancement of mediumship research.

Bill does more than just explain his journey and the process of doing high level research in this area; he places these observations in the larger context of theory and research in the field as a whole. Bill does this not only as an engineer and scientist, but as a loving father and husband whose mind and heart were transformed by what he learned.

One cannot help but be inspired by Bill's academic and personal journey. This book not only honors his children, but all children whose physical lives have been cut short. Bill shows us how their spiritual lives continue like the light from distant stars, and how we can continue to be with them on their journeys.

Gary E. Schwartz, PhD[1]

---

[1] Professor of Psychology, Medicine, Neurology, Psychiatry and Surgery, Director of the Laboratory for Advances in Consciousness and Health, The University of Arizona. Author of *The Afterlife Experiments* and *The Sacred Promise*.

CHAPTER 1

My View of Our Culture

Three things cannot be long hidden: the sun, the moon, and the truth.

Buddha

    I feel fortunate that I have been able to do a great deal of traveling in my life, particularly to foreign countries. My international travels began when I was a nineteen year old sailor in the US Navy stationed on a destroyer that took cruises to places like Cuba, Japan, Hong Kong and Formosa (now called Taiwan). After being discharged from the Navy I went to college where I earned a degree in Electrical Engineering. My working life included a good deal of traveling around the United States as well as to many countries in Europe and South America. The main reason I enjoyed my travels is that they gave me an appreciation of different cultures, and I have been a student of world cultures ever since. When I speak of culture, I'm not thinking only of the differences in clothing, eating habits, and so forth, but primarily the differences in people's cultural philosophies.

    The most significant thing that happened in my life after getting out of college was getting married and having children. That seemed to make my life complete and certainly added more meaning to it. But a huge challenge was thrown in my path—twice: surviving the loss of two sons. The first time was tough, but after being hit the second time, I became aware of what I think is lacking in our culture—and certainly was lacking in my life—an understanding of the meaning of, or reason for, our existence. I became more aware of how our society has changed from one founded on religious principles to a technological society where a subject like survival of consciousness, or in non-scientific terms, life after death, is rarely discussed in mainstream science.

Thinking about how our culture is changing made me think of how there have been paradigm shifts in societies over the centuries, especially in western society. I am reminded about people like Galileo who lived at a time when the scientists of his day believed in a geocentric theory of the universe which stated that the earth was the center of the universe and that the sun, other planets and stars revolved around the earth. Galileo's observations led him to support a heliocentric theory of the universe (first proposed by Nicolaus Copernicus) where the sun was the center about which the earth, other planets and stars revolved. When Galileo made his support of the heliocentric theory public, the rulers at the time (who were controlled by the Catholic church) decided that he could present his theory as a possibility, but not as fact, because their interpretations of religious scriptures stated that the earth could not move and therefore had to be the center of the universe. Stating otherwise was heresy. When Galileo's subsequent writings continued to defend the heliocentric theory and appeared to attack the Pope, he was punished by being placed under house arrest for the remaining years of his life.

This is the best example I know of to demonstrate how a dogmatic belief system can prevail against evidence indicating that those beliefs are wrong.

However, over the following centuries, those people who took a scientific approach to explain the physical world gradually became dominant, and in western cultures religion was kept separate. In other words, we went from a society where religious beliefs dominated to one where science dominates. Scientists have developed rigorous methods of conducting investigations, which have led to a much greater understanding of physical phenomena, but anything suggestive of a non-physical world is viewed by many scientists as heretical. In my view, in some ways mainstream science has become a dogmatic belief system.

My way of thinking of this is that the pendulum swung to the opposite side, hence the name I chose for this book.

My purpose in writing this book is to share my new perspective on a subject that suddenly became very important to me, and that I believe is important to most people: is there an existence that follows death? This question has been debated for centuries, and in general these debates have resulted in two proposed answers: 1) those expressed by traditional religious beliefs and 2) theories put forth by traditional science, theories that lack the kind of evidence I feel is necessary to give them credibility. I was not able to accept either of these views because of my need for some sort of validation.

I have spent the last twenty years searching for answers in a way that satisfied my questioning mentality. Although the answers I have found are not absolute certainties, to me they are more believable than anything I was aware of prior to my search. I have gone through an amazing transformation, one in which I have been able to fill an empty void with an understanding I never thought I would gain.

There are two general groups of people that I had in mind when writing this book:

1. People like me who suddenly lost a loved one and whose love for that person is so deep that they need something more than hope or a belief, who essentially need some sort of evidence to support the belief that they will see that person again.

2. People who have a curiosity about questions such as survival following what I call "physical death", what may follow death, and who are willing to spend time reading about the investigative work being done in this exciting new area of science.

Many people seem to have a genuine fear of death. However, a great deal of evidential information, obtained through afterlife science investigations, supports the survival-of-consciousness hypothesis and indicates that the afterlife is not something to be feared. Because these studies are different than most traditional scientific investigations, it does require the reader to have an open mind.

In addition, those who believe that there is nothing but the physical world may find this book interesting. It discusses evidential studies indicating an existence beyond the physical world, suggests that there is a "non-physical" world, and gives a glimpse of what this world may be like.

However, one thing needs to be emphasized:

Research into afterlife phenomena is a new and different type of science. Unlike traditional science, we are not just dealing with physical objects or phenomena. We are attempting to detect and evaluate non-physical phenomena.

Throughout this book I refer to the physical or material world, the world we are all familiar with that consists of organic and inorganic matter. I do this to distinguish it from a non-physical world, one that appears to consist of a "non-physical energy" that is different than any we are familiar with in our physical world.

If we were studying physical phenomena, our task would be much easier. In studies involving the physical sciences, such as disease manifestations, geologic patterns or planetary motions we have clues that, while not always obvious, are observable using the various detecting devices that science has developed such as microscopes, telescopes and infra-red sensors. Because of the non-physical nature of the phenomena involved in this new science, we must rely on methods of discovery that are quite different. These methods include using the special abilities that some people either are born with or acquire at some point in their life, many times as the result of a life threatening illness. These abilities are generally not accepted as authentic by mainstream science.

In order to show that studies in this area of science are being done by legitimate scientists, I reference the works of several professionals, mostly MDs and PhDs, in the area of afterlife research and paranormal studies, and combine this with my own personal experiences following the death of my son.

While much of the material presented here is of a scientific nature, my intent is not to present a scholarly, scientifically rigorous document. My aim is to present very unusual material in a way that can be easily read and understood by people with a variety of backgrounds, and ultimately to introduce a new paradigm through which to view consciousness, a paradigm which, like heliocentricity, requires an open mind and a shift in traditional thinking.

CHAPTER 2

The Early Days

Your present circumstances don't determine where you can go; they merely determine where you start.

Nido Qubein[2]

When I was born my parents lived in Weirton, West Virginia. My dad worked at Weirton Steel Co. which had a giant steel mill in the middle of town. The mill burned coal in the process of making steel and since there were virtually no environmental controls in those days it dumped large quantities of black soot on the town nearly every day. Most people were not aware of the effects that breathing this dirty air could have on their health.

When I was about a year old, and my brother Wayne around two and a half, my mother was diagnosed with tuberculosis (TB). As I remember my dad telling the story, he stated that in the same meeting with the doctor, when they were informed that she had TB, the doctor also told her that she was again pregnant. The only treatment for TB in those days was to put the person in a TB sanitarium so that they

---

[2] **Nido Qubein** is a businessman, motivational speaker and President of High Point University since 2005.

were isolated from the general population in order to prevent spreading of this very contagious disease. This meant that my mom was not allowed to hold nor be near her children. So my mom was put in a sanitarium in Pittsburgh, PA. when she was twenty-seven years old.

Because TB is a disease of the lungs, there was concern whether carrying a baby to full term would inhibit my mom's ability to recover from the disease, thereby raising the question of whether or not to abort the baby. The decision was made more difficult because both my mom and dad were practicing Catholics and abortion was discouraged by the church. My father told me many years later that he talked to mom's doctor and that the doctor advised him not to abort the baby because he felt my mom would recover without having an abortion.

So in 1936 my brother Lee Patrick was born at Mercy Hospital in Pittsburgh, PA. (not in the TB sanitarium) and a few months after his birth was separated from my mom, who then went back to the sanitarium. It must have been extremely difficult for her but the concern was that the baby could contract the disease. My mother had fifteen siblings, mostly girls, and ten reached adulthood. They took care of my brother Wayne and I most of the time. Lee stayed at the hospital where the nuns were able to give him the proper care. He apparently stayed there for several months, until my dad's mom, who lived in Chicago, decided to take him. That couldn't have been easy for her since she was over sixty years old at the time.

My mom died in 1937, a week before Christmas, at age twenty-nine. It was extremely hard for my dad. I have copies of letters that he wrote to his mother and stepfather during that time and my dad's frustration and sadness was overwhelming. His life, as well as his mom's, had not been easy up to that point. His dad died when he was six, he had a brother die before he was born and his sister died at age twenty eight. Now his young wife was gone and he had three little boys to take care of.

But my dad was a very strong individual. During his childhood there wasn't much money, so he worked at whatever jobs he could find in order to help his mom pay the bills. He was proud that he was able to work his way through college and earn a degree in Electrical Engineering. He taught my brothers and me the importance of both hard work and education. He also impressed on us that life can be really tough sometimes and that you need to learn how to handle it.

In 1941 dad took a job at an aluminum plant in Massena, NY, but after six months or so we moved to Chicago. We stayed with my grandmother until my dad was able to buy a house on the south side of Chicago. I had started the second grade in Weirton and finally finished it in Chicago. During that time Lee continued to live with grandma. My brother Wayne and I, who were left on our own after school until dad got home from work, occasionally managed to get into trouble of one kind or another.

One day when I was eight I started what turned into a big fire that nearly burned down some new houses that were under construction. That evening, a police officer, fireman and a person from social services appeared at our house and gave my dad two weeks to find someone to take care of us. My dad put an ad in the paper and hired a lady named Jane to watch us. She had a daughter named Pat who was about Wayne's age and they moved in with us. My brother Lee was then able to move in with us also.

In 1943 we all moved to Mount Vernon, NY. I attended Catholic schools and graduated from Blessed Sacrament High School in New Rochelle, N.Y. My dad wanted me to go to college, but I just wasn't ready. He was frustrated that the job I had at the time was selling newspapers at night to people in the bars in the surrounding towns, which kept me out most of the night. What he was not aware of is that some of the guys I was hanging out with were beginning to get involved in burglaries, something I had no interest in doing. In addition, the situation at home with Jane, who later became my stepmother, had become unbearable and on top of that, I was a prime candidate to be drafted into the Army. I decided that the best solution was for me to enlist in the Navy, which I did in 1953. My dad was not happy about it, but I consider that decision one of the best I've ever made. It enabled me to get a fresh start in life. The Navy sent me to Electricians Mate (EM) school and following that I was stationed on a destroyer.

I was discharged from the Navy at the end of 1957, and the following year I enrolled at the University of Colorado (CU) where I studied Electrical Engineering. An integral part of my studies was learning what is generally known as the scientific method, which is defined by the *Oxford English Dictionary* as *"a method or procedure that has characterized natural science since the 17th century, consisting in systematic observation, measurement, and experiment, and the formulation, testing, and modification of hypotheses."* Learning the rigid objectivity required by this discipline had a strong influence on how I viewed our world in subsequent years.

In 1963 I graduated from CU and moved to Northern California. Finally I was able to relax a little and begin enjoying life. In addition to having a good job, I went snow and water skiing, learned and later taught Jiu Jitsu, did some skydiving, got a pilot's license and sometimes just hung out with friends. I also spent a lot of time with a young lady I met where I worked named Diane, who had a little boy named Jim from a previous marriage. Jimmy was almost three at the time and he became my little buddy. When I was quite young, I remember thinking that some day I wanted to have

The Early Days

children. For some reason that seemed like something that I would enjoy.

Diane and I were married in 1965. In 1967 our first son Ricky was born. I was *elated*. He appeared to be a healthy little guy, but during the post-birth examination our doctor detected a problem and told us he wanted to have some tests done. The very next day he informed us that Ricky had a serious heart defect called Truncus Arteriosus. His pulmonary artery and aorta had failed to separate into two vessels. This meant that freshly oxygenated blood was being mixed with de-oxygenated blood, causing a reduced level of oxygen in his blood stream as well as the accumulation of fluid in his lungs, making it difficult for him to breathe. He was placed in the Neonatal Intensive Care Unit (NICU) at Stanford Hospital, but there were no surgical procedures for his defect in those days. All we could do was wait and hope for the best. After a short time we asked if we could take Ricky home. The doctor agreed, but told us we would need to suction the fluid from his lungs. A nurse then handed me a "turkey baster" with a piece of tubing stuck on the end as the suctioning device. That not only surprised but also concerned me because I wasn't sure how we could control the amount of suctioning, so the next day at work a friend and I built a small motor driven suctioning device with a control valve that allowed much safer suctioning of Ricky's lungs. That experience planted the seed in my mind that someday I wanted to start a company to develop products for the medical field, which I finally did about ten years later.

But our happiness at having Ricky home was short-lived. There were times when he seemed to be improving, but they were followed by times when it was obvious he was really struggling just to breathe. Then one night Diane woke me in the middle of the night because she sensed that he had stopped breathing, We were able to get him breathing again and rushed him to the hospital in the middle of the night.

Several days later, Ricky died in the hospital. It was by far the most difficult thing Diane and I had ever experienced. It took awhile for us to move beyond our grief but we were gradually able to get on with our lives. We were young and felt confident that we could have more children, and later that year I adopted Jimmy.

Two years later our son John was born, followed over the next few years by our daughters Becky and Angie. Because of our experience with Ricky, there was a little concern when John was born, but he was a healthy, happy baby, and grew up to be a wonderful young man. Life was good. It was wonderful to be able to watch my kids grow and learn, to go on vacations together, attend their many sporting events, go on "dates" with my daughters and to just enjoy life with each other. John and I spent a lot of time together and often talked about what we were going to do together in the future.

Then one evening I received a phone call telling me that *John, who was twenty-two years old, had been killed in a motorcycle accident.*

This time I was not able to just "get on" with my life. I had gotten really close to John and the pain and desperation I felt can not be put into words. Human language is far too inadequate to describe my feelings. I began a search to try to find out what is the meaning to what we call "life" and in particular to try to find out if I would ever be with my son John again.

One of the things I noticed at our house following the memorial service for John was that some people were really uncomfortable trying to console us, and I found myself trying to console these people. However, I understood because I had difficulty finding words that I thought would be comforting to friends who had recently lost their daughter.

Diane and I agreed that it helped us to talk about John, and to hear stories about him, especially from his brother, sisters and friends. All we had left were memories, and they were suddenly truly important. However, I soon began to realize that talking about the loss of our son sometimes created an awkward situation. For example, when meeting new people, they often ask "How many children do you have?" At first, after losing John, I wasn't sure what to say. I found it was easier to just say three, but to me that was denying the existence of John, something I just could not do, so I began to answer four. Then they might ask something like "What are they doing now?" At this point, I would tell about the others and then explain losing John. I began to notice that if I mentioned John to some of our friends, or told people who didn't know him about our loss, they became noticeably uncomfortable. Some people would immediately try to change the subject. My first reaction to this was a feeling of resentment. But I know that people often have difficulty dealing with their own emotions, and it can be equally difficult for them when hearing about a loss that others have suffered. I also came to realize that some people really do have a fear of death, and are uncomfortable when the subject is brought up.

It occurred to me that one possible cause of the reaction I was seeing in these people could be our western cultural upbringing. Our culture is now based largely on technology where we can control most of the things in our lives. Although there are still millions of people who are religious, our society is slowly getting away from traditional religious beliefs related to what follows death, and often there is nothing to replace these beliefs. I began to realize that one of the unintended results of the increasing influence that science is having on our culture is that it has essentially removed from many people's daily lives any reference to a hereafter. This approach to viewing what we think of as reality has, in effect, instilled in many people a belief that there is no existence beyond death.

CHAPTER 3

Time For A Change

We cannot solve our problems with the same thinking we used when we created them.

Albert Einstein

When I was in college, in addition to studying for an engineering degree, I was also studying for a minor in business, so most of the non-engineering classes I took were business classes. The few electives I chose were science classes such as physical anthropology and astronomy. I never even considered something like philosophy. I had no idea what the courses in liberal arts were about, but I had an attitude that said they were probably mostly useless courses with no real-world applications. As a result, I had no idea what was going on in this area of study, and was completely unaware that it included theories of what lies behind our existence. I do not recall ever having a serious discussion regarding "what life is about" with any of my fellow students, or for that matter, with any of my engineering or business colleagues later in life. I now realize that, in general, most of us were unaware of how a relatively new science—that of philosophy—was advancing in parallel with the advancement of the physical sciences. It was not until after I lost John that I became aware of what was missing in my education.

I had to read many books about subjects that are not a part of mainstream science before I learned of the changes that have occurred in Western Culture.

There have been many paradigm shifting events in history, such as the use of fire to cook food, the invention of the wheel, the domestication of animals, and so forth. My personal favorites are those caused by people who make observations that are in conflict with an accepted existing paradigm—people who were not afraid to present these "conflicting observations" to their fellow man—people like Galileo. The reason Galileo is one of my favorites is because he spent a good part of his life trying to open the minds of the people of his day, especially the "educated" ones, and was essentially persecuted for his efforts—by those very same "educated" folks.

The question in my mind is this: How can you get people to think differently than they have been taught, usually from a very young age? There is a lot of talk these days about "thinking out of the box" which to me means to think on your own instead of blindly following the teachings and doings of others. This raises the question "why is this important?" In trying to answer this for myself, I concluded that most of the significant advances in human culture have been brought about by people who thought of, and implemented in one way or another, changes that have solved problems that humanity was faced with. This can be seen in the advances made in agricultural methods, health care, much faster and more convenient modes of transportation, methods of communication, conveniences in our homes, and many, many more.

In addition, because of people like Copernicus, Galileo, Kepler, Newton, Darwin, Einstein and countless others, we not only have a much better understanding of the physical world, but we are taking the first steps that will allow us to venture from our planet for extended periods of time. And yet, I find that we as a culture are in some ways still in the dark ages.

For the first fifty-six years of my existence, I was quite comfortable with the life I had. I had been educated in some excellent schools and accepted that, through the use of science and what is known as the scientific method of investigation, we could and would eventually find the answers to most, if not all, of the questions that have been intriguing man—probably at least since we evolved from our caveman days. As far as the spiritual side of things was concerned, for some reason I was just not able to accept that we could find all of the answers in religious scriptures. I understood the general teachings—concepts such as the golden rule—but as a very young man I began to question and eventually reject many of the religious tenets as presented to me by my teachers. The material they presented didn't seem to be based on any sort of factual evidence. Either you believed it, or not.

However, there are two things that I remember from my first grade religion class: 1) a really cute little girl that I had a crush on and 2) being asked the questions "Where do we come from?", "Why are we here?" and "Where do we go when we die?". I was intrigued by those questions and never forgot them. As the years went by I gradually came to believe that by applying a sufficient amount of scientific analysis to the subject the answers would eventually be found. I was sure this would not occur in my lifetime, but that didn't bother me. I wasn't really all that concerned about the answers to those questions anyway, even after my son Ricky died. Following a period of grieving I was able to go on with my life. I did not give much thought to whether he survived the death of his body, what he might be doing now if he had survived, or other similar thoughts. In fact, I essentially still thought of us as only our physical body.

But on May 2, 1991 when John was killed, the comfortable little cocoon that I was in completely disintegrated. I could no longer just ignore the questions about survival after death. I had to *"know"* the answers. At that time I was solidly entrenched in the world of science, and felt that science was the only hope I had of finding these answers. All of a sudden I felt a tremendous sense of urgency. My "belief system"—that science could provide the very badly needed answers for me—was failing miserably. It became clear to me that I would need to embark on a journey very different than any I had ever taken.

As I began to search for answers I became more aware that the result of my having given up on traditional religion—combined with the influence that my science-based education was having on me—is that I had absolutely no vision of what happens after we die. There was no place I could turn to that would help me find answers to all of the questions that were suddenly flooding my mind. It's hard to describe the feeling of hopelessness I felt in those first months following John's death. When Ricky died, I fought back the tears and refused to cry, even when Diane and I stood at the graveside watching them put the box containing his little body in the ground. Men aren't supposed to cry. But when I lost John, I had no control over the sudden episodes of breaking down that went on for months. It didn't matter where I was or what I was doing. There were times when I had to pull my car off the road, or leave the room when with others. I cannot describe the sadness and despair that sometimes simply overwhelmed me. I desperately wanted to know if my son had somehow survived the death of his body, but I didn't even know where to look for the answers. All I knew was that I *had* to find them.

CHAPTER 4

The Near Death Experience

Faith is taking the first step even when you don't see the whole staircase.

Martin Luther King, Jr.

The night John died, I made a promise to myself. I swore that I would spend the rest of my life trying to determine where my wonderful son had gone. I could not accept that he no longer existed. I had no idea how I was going to begin my search, but the next day a friend who had recently lost her daughter gave me a book that began a dramatic change in my thinking that is still going on more than twenty years later. That small paperback book, titled *Life After Life* by Raymond Moody, Jr. M.D. was the beginning of a reading binge that continues to this day, although at a less furious pace. To summarize the content of this book, Moody, who holds both MD and PhD degrees, noticed that many people who had come very close to death, usually as the result of an accident or some other sudden brush with death such as a heart attack, described what they felt was an experience similar to what they believe will happen when we die.

These experiences typically included several elements: an apparent "leaving of their body", traveling through a "tunnel", being met by deceased people—some of whom they recognized as deceased loved ones—feeling the presence of a "being of light", having a life review and then returning to their physical body. One of the fascinating things is that, following recovery, these people could relate things going on in the hospital, or occasionally elsewhere, while they were clinically dead! For example, they could describe what the doctors were doing to their body while trying to resuscitate them, or sometimes tell what was going on in a different area of the hospital at that exact same time. This occurred during the "out-of-body" portion of their experience. Dr. Moody termed this a Near Death Experience (NDE), but the key point is that they didn't *die*, they just left their body for a brief period of time, and then returned! Needless to say, reading that these people encountered deceased loved ones gave me hope that my son John did survive the death of his physical body.

After I read Dr. Moody's book, I began looking for other books on the subject, and soon found one entitled *Closer To The Light* by Melvin Morse M.D. Dr. Morse is a pediatrician who learned about near-death experiences when a young girl named Katie, who had nearly drowned, and had been in a coma for three days, told Dr. Morse a very unusual story about what had occurred while she was in coma. The story she told was very similar to those related in Dr. Moody's book. A key difference is that this case involved a child. Skeptics could argue that adults may have heard of NDE's and may have fabricated their experiences, but as Dr. Morse points out, a child who survived a nearly fatal accident is not likely to have heard of an NDE nor to fabricate such a story. He decided that studying children would give him an opportunity to deal with a pure population, and decided to do a study of the phenomenon of NDE's among children.

I am going to paraphrase some of the material from Dr. Morse's book.

When searching the medical literature for information about the experience of his patient Katie, Dr. Morse could find very little mention of the near-death experience. This was a concern because the effects of this experience were having a profound, positive effect on his patient, and yet this phenomenon was apparently being ignored by mainstream medicine. The only information he could find regarding NDE's was in Dr. Moody's book, which was outside of mainstream medicine. An additional observation made by Dr. Morse is that any mention of this phenomenon to many of his colleagues was greeted with a high degree of skepticism. They typically automatically rejected the validity of these experiences and assumed that his patient was hallucinating from the drugs she was being given.

Some of them even suggested that Dr. Morse was hallucinating and "took to whistling the theme to *The Twilight Zone*" whenever he would bring up the subject.

Many of his medical colleagues thought that near-death experiences shouldn't be dignified by scientific investigation. Dr. Morse expresses the opinion that he believes medical science has tried to ignore the near death experience because it raises the question of whether there is life after death, and that this question defies the rigid objectivity "hammered into us in medical school".

It is 2012 as I write this and Dr. Morse's book was published in 1990, yet I don't think there has been much change in this attitude within the medical (or scientific community in general) since that time.

In spite of the discouragement, Dr. Morse went ahead with his study. To test the hypothesis that a person must actually be "near death" for the phenomenon to occur, he assembled two groups of children. The first group served as the control group and consisted of 121 children. These children were critically ill, but not near death. They were on artificial lung machines, were treated with tranquilizers and narcotics, ranged in age from three to sixteen years of age and were hospitalized in the intensive care unit at Children's Hospital. All of them had been bedridden for a long period of time, and most had been heavily medicated at one time or another. They had been treated with the same drugs that some doctors thought had created Katie's visions. All 121 children in the control group had serious diseases, but had less than a five percent chance of dying.

The second group was the study group and consisted of twelve children who had come close to death and who would probably have died or faced severe handicaps had it not been for modern medical care.

Dr. Morse did not mention or hint at any of the elements of an NDE when questioning any of the children.

The results were as follows:

None of the 121 seriously ill children had anything resembling a near-death experience.

However, most of the children in the study group—those children who had survived cardiac arrest, returned from deep comas and so forth—related at least one of the NDE characteristics. They described being "out of their body", traveling through some sort of tunnel, seeing a light, being visited by deceased people, seeing a Being of Light and maybe even deciding to return to their bodies.

Dr. Morse then interviewed thirty-seven children who had been treated with almost every kind of mind-altering medication known to pharmacology. These children had been given anesthetic agents, narcotics, Valium, Thorazine, Haldol, Dilantin, antidepressants, mood elevators, and painkillers. He wanted to see if these drugs could be the cause of NDEs as many of his colleagues believed.

None of these children had anything resembling an NDE

What I find interesting is that, even though it is estimated that millions of people worldwide have had NDE's, I had never heard about it prior to losing John. When mentioning it to friends and acquaintances, I received a look of disbelief from some of them who appeared incredulous that I could actually believe such outrageous claims. But the other thing that I found interesting is that there were a significant number of people who had not only heard about this phenomenon but also believed it. A couple of them had even had such an experience themselves. However, because of the taboo placed on such an "occult" subject, they didn't dare mention it to most people for fear of being ridiculed.

As I began to understand more about the subject of the afterlife, I noticed that if I attempted to tell people about my newfound knowledge, I would get this look from some of them that said essentially "Poor Bill—losing his son has apparently caused him to lose sight of reality." I didn't like that look, but in addition I was frustrated because I could not come up with a way to explain to these people that not only was I still normal but I was also beginning to learn about evidence supporting the hypothesis of survival of consciousness—in other words—what may actually happen when we die. This frustration, coupled with a strong desire to understand as much as possible about what lies behind our existence, made me determined to continue my search.

CHAPTER 5

A Glimpse of Other Paranormal Phenomena

It is the mark of an educated mind to be able to entertain a thought without accepting it.

Aristotle

In the early days following John's death, everyone in our family was searching for ways to get back to something resembling a normal life. Each of us had our own ways of trying to cope with what was a huge loss for all of us. Diane began working in the NICU at Packard Children's Hospital. Jim was working at a large biotech company, and continued with his favorite sport—running—as well as spending time with his friends. Becky was in college at UC Santa Barbara where she kept busy with her schoolwork, working at the Santa Barbara airport and spending time with Aaron, her future husband.

Angie was in high school, and in addition to keeping up with her schoolwork, was MVP in soccer and tennis her senior year. She then went off to the University of Colorado with her best friend Amanda. I was happy that she had such a close friend with her. They were roommates, sorority sisters, went skiing together and did all the things that young college girls do. Then suddenly Amanda was killed at the end of their first semester in a tragic accident. Again, I can not describe how difficult it was watching my young daughter struggle all over again just to keep on going, nor can I begin to describe the pain that she and Amanda's parents were going through.

Beginning with Moody's book, I was gradually being introduced to the idea that there is more to life than just the physical world, and that there were people who were normal in every way except that they apparently had a sensitivity to things that most of us are not aware of. I have since learned that there are many of these people, but most of them are not known to the general public, often due to the fact that our culture does not accept them as genuine.

My personal introduction to paranormal phenomena began in a rather subtle manner. I did not actively seek to experience them, but they were sort of serendipitously shown to me. I am going to give examples of a few incidents that happened in the early years following John's death. These incidents are examples of phenomena that occur quite often, but that I was barely aware of. The first is a psychic reading that Diane describes in notes that she kept.

Psychic Reading from Pat McAnaney.—June, 1991

"We received a phone call from a friend. She told us that an acquaintance was staying with her that weekend who happened to be a psychic medium. She said that he was available at her home to give Bill and me a reading if we were interested and we would not be charged. Neither of us had ever met or gone to a psychic and therefore had no prior impressions or judgment regarding them. In fact, we had only heard about fortune tellers, and that they were kind of wacko and probably bogus. We decided that we had nothing to lose

A Glimpse of Other Paranormal Phenomena

and that anything that would help our grief at this time might be worthwhile.

Pat McAnaney is known as a clairvoyant, apparently born with the ability to see and feel beyond the range of the normal five senses. He said that this innate gift allows him to tune in and read the emotional energy of others. This energy is known as color auras.

(Wikipedia states that, in a paranormal sense, an aura is "a field of luminous multicolored radiation around a person or object.")[3]

On May 18, we went to our friend's house. Bill and I sat down together in a room with Pat with open and curious minds. Initially he read our "auras" and one of the things Bill remembers is that he was told that he had a strong aura, so was an old soul who had lived many lives. Pat then said:

"You lost a son recently, in an accident." He went on to say that our son left suddenly and when he realized what had happened he thought 'Damn—this is not good. I'm in trouble'. Then, Pat added, John 'said' 'Oh well! Let's see what's next'—and he was ready to go on to this new adventure, quickly shrugging off any regrets. Kind of like 'it happened—nothing I can do about it now—so let's go!' John was like that—always excited about whatever adventure was coming next—where would he go, who would he meet, what would he learn?"

---

[3] I reference Wikipedia several times throughout this book. This website is written collaboratively by largely anonymous Internet volunteers. Anyone with Internet access can write and make changes to Wikipedia articles. For this reason, when I quote something from Wikipedia that may have more than one interpretation, I specify the meaning I intend in the Glossary at the end of this book.

Diane then explains how she and John used to have talks about why we are here, where we go, etc. and felt that the reading was real.

Her notes continue:

My Phone Call to Pat McAnaney—June 11, 1991.

"I was feeling so lost, sad and hopeless and wondering how to go on.

I called Pat McAnaney. I just knew that I was desperately needing 'something' to help me, so I asked Pat if perhaps he felt or heard anything from my John. Pat simply replied that I would be receiving a very special letter within the next few days—and that this letter was important."

The Following Day—June 12, 1991

"Sometime during the afternoon I went to the mailbox. There was a letter to us from C. P. This young man, who was a friend of John's, was responsible for John's death on May 2, as he was the driver of the motorcycle, under the influence, and went out of control—".

In her notes, Diane goes on to explain her feelings at that moment.

When offering the "sitting" with Pat, our friend had told us that she did not say anything to him about John. At the time, I had no idea that there were people who had an ability that somehow involved the use of a sixth sense that allowed them to "see" things that most of us can not see. In fact, the reason I had to use Diane's notes to relate this story is that, because the experience was so foreign to anything

I had ever even heard of, I basically forgot everything that happened with Pat except for two things:

1. His comment that I was an old soul. I wasn't sure what that meant, other than that I had lived many previous lives, but I took it as a compliment. Don't ask me why—it's probably an ego thing.

2. That he predicted we would receive an important letter within the next few days, and that it actually happened *the next day*. I could not ignore that. There was no way he could have known the young man responsible for our son's death was going to send us a letter when he did. If his prediction had been more general, I would not have been so impressed, but he specifically predicted a letter, it did arrive within a few days, and it certainly was important. I tucked that piece of information away somewhere in my mind, forgot about it and went on with my life as best I could, reading and searching for a "scientific" answer to where my son had gone. In spite of being given a glimpse of mediumship, my supposed scientific objectivity would not allow me to accept it on face value. I needed to have it shown to me in a proof-based way.

My attempt to have an Out of Body Experience (OBE).

About eighteen months after John's death, while reading books on various paranormal phenomena, I noticed a reference to a book entitled *Journeys Out of the Body* by Robert A. Monroe, a person who claimed that he was able to leave his physical body almost at will. Since I was aware that leaving the body was listed as one of the elements of a Near Death Experience, I decided to buy the book and read it. After several chapters of background material, Monroe describes the method he used to accomplish this "leaving of the body." The process essentially consisted of four steps:

1. Achieving a state of complete relaxation.

2. Reaching a point where the body experiences a "state of vibration".

3. Learning how to control the vibrations.

4. The separation process.

In order to assist people who were interested in learning this process, Monroe had founded The Monroe Institute, located in Virginia. After reading the book I decided to contact the Institute. I was informed that a person could register for a training session there and would be taught how to induce one of these experiences. As an alternative, they could buy a set of audio tapes that offered the same training and could attempt to accomplish the "separation" on their own. I was intrigued by the idea so I ordered the tapes.

The tapes consisted of instructions and soothing sounds such as ocean surf that were intended to help you relax—the key first step in the process. There were six tapes that you listened to sequentially. This was typically done at a time when you were not extremely tired so that you could relax without falling asleep. The idea is to get to a place where you are on the edge between being asleep and awake. I wasn't very optimistic about my chances of accomplishing this because I had become aware of my inability to relax my mind the few times I had tried to meditate. However, I forged ahead, found an old Sony Walkman tape player and prepared for a new experience.

I listened to one of the tapes each night when I went to bed for a week or so and found that they were an excellent way to lull me to sleep, but nothing more. Then one night, while listening again to the sounds of surf crashing against the shore, I suddenly felt my body begin to vibrate! I became really excited and was ready to pursue this new adventure, but after several seconds, the vibrating stopped. I was really disappointed but determined to keep on trying. However, over the following weeks there was only one more time that I began to have the same experience, and I never achieved the success I did that first time. I could not get myself to that third phase where you maintain and control the vibrations, so I did not have a successful OBE, although it appears that I came close. After awhile I tired of trying and gave up the effort. I recently began to read this book again and from what I have read so far it appears that I gave up too easily. I'm thinking it would be fun to try again, especially since I now have a much lower level of skepticism.

My next exposure to what is classified as paranormal is called a Past Life Regression

"Past life regression is a technique that uses hypnosis to recover what practitioners believe are memories of past lives or incarnations, though others regard them as fantasies or delusions. Past life regression is typically undertaken either in pursuit of a spiritual experience, or in a psychotherapeutic setting. Most advocates loosely adhere to beliefs about reincarnation though religious traditions that incorporate reincarnation generally do not include the idea of repressed memories of past lives." Wikipedia

I had never heard of past life regressions until I began reading books related to paranormal phenomena. If you continue to read on Wikipedia where I left off above, you will see that there is still a great deal of skepticism about this subject, the same as with most other paranormal phenomena. My only exposure to hypnotism was a show I attended years earlier where people were invited on stage, hypnotized by the entertainer (who was a magician, not a therapist) and then, following suggestions from him, began doing the silly but funny things he was suggesting. I figured the whole show was probably set up ahead of time. I had heard that hypnotism was used as a therapeutic tool, but I had no interest in trying it. Not long after losing John, Diane asked me if I wanted to read a book about past life regressions that she had read but I told her no. It just sounded a little too far out for me at the time.

Then one day, a few years after losing John, I discovered that one of the top people in the field, Edith Fiore, PhD, lived in a town near me and my curiosity got the better of me. I made an appointment and went to her office. I had absolutely no idea what to expect, and in fact had no expectations, but was determined to have an open mind. I sat in the easy chair in her office while she slowly hypnotized me, which was also a new experience for me, and which I learned was merely a method of causing a person to be completely relaxed and to block out peripheral activity in order to permit you to concentrate.

She then asked me to describe what I was "seeing" in my mind, but did not "suggest" anything. She just listened. I was a little surprised when I began to get a vision in which I was standing at the bottom of a long, steep hill in what appeared to me to be an ancient city. It was a clean city, and there were white, two or three story buildings lining the far side of the street from where I was standing that went up the hill to the left. (I had never visited a place similar to what I was seeing in my vision, but I recently saw a photo of Santorini, Greece that reminds me of the city I saw in my vision. I have never been to Greece, but it is now on my list of places to go).

I walked to the top of the hill and then began to walk along a tree-lined pathway alongside of the road, similar to a sidewalk. As I walked along I passed a couple of middle aged women walking the other way who were wearing colorful, full-length dresses. I continued walking until I came to an arched entry to what was apparently a park. I entered the park and saw some children playing at the far end of the park, running and laughing and playing some sort of game, so I walked toward the children. As I approached, one of the children, a boy who appeared to be about seven or eight years old, ran toward me. It was John! He walked over to me and asked me to come over and play with him and his friends. I told him that I would like to, but that I had too much to do. Then the vision faded.

Dr. Fiore had given me a copy of the tape recording she made that day which I have not been able to locate, so I am repeating this from memory almost twenty years later. I am surprised as I write this at how clear the images still are. The entire vision probably lasted around twenty minutes. I asked Dr. Fiore how I did, and she said that, on a scale of one to ten (ten being the best) she would give me a nine. I remember one thing in particular about that day—as I left her office I felt really relaxed and pleasant. I made an appointment a few weeks later for another regression, but I had trouble relaxing, and did not have as good a vision. I don't even remember it at all, and soon after I essentially forgot about the subject.

The comment about me being too busy to play with John at the end of the regression ties in with something that happened when he was about fourteen, and that had a profound effect on our relationship, an effect that I am very grateful for. One day I walked into the kitchen and heard John and his sister Becky having an argument. John was apparently pretty frustrated. As with most siblings, John and Becky used to argue, but this one seemed larger than normal. John appeared to be really angry and I felt like he needed to get some anger out so I told him that if it would help, go punch the wall or something. I had never made this suggestion in the past. So John walked to the other side of the kitchen, took a swing at the wall, and punched a hole in it! It caught me completely by surprise. I let him calm down for a minute and then asked him what the problem was. "Dad, you're always so busy you never have time to do anything with us". I knew immediately he was right. I had ended up doing the very thing I had promised myself I would never do—get so busy that I wouldn't have time for my family. I was working long hours at the company I had started, doing a significant amount of the work on our house remodel, and trying to raise a family with four kids. I told John he was right and that I was going to change that—and I did.

Soon after, he and I did one of the things he had wanted to do for quite awhile. We started at the creek by our house and walked a couple of miles upstream in the creek, then sat down, ate our apples and talked about things—things John had been wanting to talk about—things like what he wanted to do with his life, what he and I might do together in the future and so forth. That was the first of regular one-on-one times I spent with each of the kids. Work was still important, but now it was second to my family. Over the next eight years, John and I took many trips together. This included trips around the US where we camped out and went to places like the Grand Canyon, Mesa Verde in Colorado, took a trip to Hawaii to visit my dad or just went to the gold country to go fishing or pan for gold. We also traveled to Italy where we spent time with my relatives there, climbed the leaning tower of Pisa, rode the overnight train from there to Paris, and so on. Sometimes we just hung out together. The summer before John was killed, I rented a large houseboat on Lake Shasta and told each of the kids to invite a couple of friends. My brother Lee showed up with his family plus a couple of teenage girls from Italy who were staying with them for the summer. That was one of the most fun filled week-long vacations we had ever had. Now that John is gone, those memories are truly important.

Another Psychic Reading

One beautiful summer afternoon around the year 2001 I was sitting in the outside picnic table area at our local beer and burger place with a large group of friends. A young lady named Ellen who I had not met but who was part of the group came over and, without even an introduction, asked me if I was a shaman. I figured she must be a new-age person or something and told her no, but then asked her why she thought I might be. She said that she had psychic abilities and could see "spirits" near me. I was intrigued by this and so we talked for awhile. I told her about losing John, and asked her if she could do a reading for me. She said "sure" so I set up a meeting with her.

When I arrived at her place a couple of days later, as I walked in she immediately said "I had a vision yesterday about 2 pm, and I almost called you. I saw you falling down a dirt hill by a big tree and a young man was standing there laughing at you." She described the young man and the description fit John, who she had never met. I was definitely surprised. The day before, around that time, I was in my yard down by the creek (by myself, I thought) picking up some logs I had split. A couple had fallen down the embankment and as I climbed down to pick them up I slipped, fell down the embankment, and ended up on my back in the creek. I have to admit I was impressed when Ellen described what had happened. I joked with her about how John was probably laughing because he pushed me.

I don't recall anything else of significance that came from that session. Sure, I had told her about John, so her describing him could be explained, but I could not ignore the fact that she was able to tell me about an event that had just happened and that she could not have known about. And yet—I still had my doubts. Not long after that session she moved back east, and I went back to reading my books.

So up to this point, my personal experience with the paranormal consisted of two brief sittings with psychics, an attempt to have an out of body experience and one past life regression, spread out over ten years. In spite of these glimpses of phenomena very different than anything I had ever experienced, I was still looking for something more solid—something that met the scientific criteria that had been ingrained in me as necessary before it could be accepted as fact.

After about ten years of reading dozens of books that included many on NDE's, several written by psychics, books such as *The Holographic Universe* by Michael Talbot, *The Matter Myth* by Paul Davies and John Gribbin, *A Brief History of Time* by Stephen Hawking, books on dream analysis, out-of-body experiences, and so forth, I still had only a very sketchy picture of what life after we die, assuming there is such a thing, might be like. Most of the books written by scientists only talked about the physical properties of the universe.

While the people who had experienced an NDE described their experiences, it didn't shed much light on what the afterlife is really like, and from what I could see only made the case that one does exist. What also made it difficult for me to accept as fact was that many of these people had what I considered a very strong religious bias in reporting their experience. That didn't make it wrong or even inaccurate, but simply more difficult for me to accept because of my strong scientific bias. Another inhibiting factor is that I couldn't relate to something like an NDE because I'd had no personal experience with it—I had never "nearly died."

When I became personally involved in the subject of the next chapter—the scientific study of mediumship—my perceptions and my attitude changed dramatically. At first I was skeptical, but my skepticism about this phenomenon eventually completely dissipated.

CHAPTER 6

The Study of Mediumship

To study the abnormal is the best way of understanding the normal.

William James[4]

During the first ten years following John's death, I was gradually able to get back to a normal routine. One of the things that helped a great deal was having good friends. For the first few years I continued working at Paramed Technology, the medical products company I founded in 1978, but although I was trying hard, it was difficult to stop thinking about John and focus strictly on the job, especially in the days right after he died.

I remember talking with a colleague at work who was frustrated that I was having this difficulty. He said to me something like "It's been more than five weeks—you should be over it by now!" He then told me that if his son were killed, he would just get rid of any reminders of his son. He

---

[4] **William James** (January 11, 1842-August 26, 1910) was an American philosopher and psychologist who had trained as a physician. He was the first educator to offer a psychology course in the United States.

knew that my son had been killed in a motorcycle accident, so I asked him how he thought I should get rid of all the motorcycles in the world. He went very silent. I think this little incident points out just how difficult it is for many people to understand how devastating the loss of a child is.

In 1994, I left Paramed and began traveling to South America where I set up and managed product distribution for medical product companies. My initial territory was Brazil, Argentina and Uruguay, and I later expanded into other countries in the region. This turned out to be a wonderful experience. One of the things I noticed was that if I mentioned the loss of my son, it did not seem to make people feel uncomfortable. They appeared to me to be very compassionate, and more accepting of death. I continued this work until around 2001, but quit not long before the terrorist attacks of 9/11 occurred.

One evening around August or September, 2002 I was sitting in our living room with the TV on flipping through the channels looking for something of interest and I stopped at a show called "The O'Reilly Factor". The host, a guy named Bill O'Reilly, was saying "So folks, one more time before we sign off for the night, my guest is Gary Schwartz and the name of his book is *The Afterlife Experiments—Breakthrough Scientific Evidence of Life after Death*." I ran to the kitchen, grabbed a pencil and paper, and wrote down the author's name and the name of his book. The next day I ordered a copy from Amazon. Diane and I were leaving on a trip to meet my brother Wayne and his wife Darline in Utah in a few days and I wanted to read it on the trip.

Fortunately it arrived before we left, and it was exactly what I was looking for. Gary Schwartz, PhD is a professor of psychology, medicine, neurology, psychiatry and surgery at the University of Arizona. He received his doctorate from Harvard and taught at Yale before moving to the University of Arizona where he was conducting afterlife research. He had been studying and testing people called "mediums" who claim to be able to receive communications from deceased people, and the book was a publication of his results to date. In his book, Dr. Schwartz described the experiments he performed with the mediums and explained how he evaluated their performance.

A typical evaluation was conducted as follows: A medium would sit in a room at the lab, and a "sitter"—the person who wants to receive a message from their deceased loved one—sits in a separate area of the room. The medium, who was given just the first name of the deceased loved one, would begin making statements based on what he/she was perceiving. There is no visual contact between the medium and the sitter, and the only verbal contact is the sitter's response to statements made by the medium. The sitter confirms or denies the validity of the statement with a simple yes or no reply—no further help is permitted by the sitter. The medium's comments, which were recorded, were later scored by the sitter using a +3 to –3 system, with +3 being most accurate and –3 being false (as far as the sitter knows). The accuracy of the information received by the mediums in these experiments was astonishing. Things that no one would normally know about the personal lives of the sitters were revealed with surprising accuracy.

I decided to contact Dr. Schwartz. My first attempt was a letter. After a short time I did not receive a reply to my letter so I found Dr. Schwartz' email address and sent him an email. This time I had better luck. I told him of my interest in his work and also that I was interested in being involved in similar experiments. It turns out that he was going to be in San Francisco in early November for a meeting, and since I live near there we agreed to meet at his hotel. We met for breakfast and that was the beginning of a long and very interesting relationship between us. After asking him questions about his research methodology, I told Dr. Schwartz that I would like to in some way try to support his work. I invited him to dinner at our house that evening and was excited when he agreed to come. After dinner Dr. Schwartz told me that he was going to send an email to a few of the test mediums involved in his research to "see what they get." He didn't tell them where he was or who he was with. He only gave them the first name of my son, John. The three mediums were Janet Mayer, Allison DuBois and Mary Ochino.

The next morning after returning home, Dr. Schwartz received replies from the mediums and forwarded a total of nine readings (three from each medium) for me to evaluate using the scoring criteria explained in his book. At first, as I began reading the material, I was a little confused, and perhaps even disappointed because it was not what I had expected. Like many skeptics, I expected the conversation to be the same as if the mediums were simply talking to my son. I didn't understand that they receive images, hear voices or sounds or even feel something physical, such as a pain in their chest, and must then try to describe what they are sensing. What really had me intrigued was that three different mediums, in three different states, who had no idea who I was, who knew nothing about me, and who had no time to try to research me even if Dr. Schwartz had somehow secretly given information to them about who I was, could come up with information unique to me and my family. *By internet!* It took quite a while before I finally grasped the only way that this could be possible.

A Typical Reading

I think it will be helpful to the reader if I show what an actual reading is like. Before introducing you to the first reading, I want to explain the basics of evaluating a reading. As the medium receives each image, voice message or feeling (what I call "perceptions"), they can only state what they are perceiving. Some of these individual statements will often apply to many people. Sometimes an image may not be clear, or a sound may be difficult to interpret. I think of these situations as analogous to a radio, TV or cell phone with poor reception. Other times the perception may be clear, but as far as the sitter can tell, it does not fit their situation. This statement is counted as a miss. Statements may be true but not highly significant, so would not receive the highest rating. The score for any particular reading is determined by the number of statements that fit the sitter's situation.

The more true, unique and meaningful statements made by the medium, the higher the score.

The first reading is one from Janet Mayer in which my son John came through. The reading is reproduced as it was originally received and then commented on by my wife and I. Our comments in response to each of Janet's statements are in all capital letters. This reading is a good example of what medium's do, and I urge you to read it with an open mind. It should be noted that all of the following samples of readings, except an impromptu one by Gordon Smith mentioned in a later chapter, were email readings being done as part of the experimental work Dr. Schwartz was conducting at the time.

Janet labels this reading "A Specific Person From Susy." Susy is a reference to Susy Smith, a medium who Dr. Schwartz had met before she died. She told him that after her death she was going to "assist" him in his mediumship studies. In his book *The Afterlife Experiments* Dr. Schwartz tells the story of how he first met Susy when he began his work at The University of Arizona. Apparently she was quite a character. Following her death, Dr. Schwartz experienced some very unusual phenomena while working with other mediums which indicated that indeed Susy was continuing to assist him in his studies of mediumship. Susy was now a "deceased person working with other deceased people" so Dr. Schwartz labeled this the "double-deceased paradigm."

Dr. Schwartz describes in detail how this all came about in his book *The Sacred Promise.*

In the reading, Janet is typing what she is experiencing as it happens, so there are a few minor typing errors. The name "Gary" in this reading or in other places in this book refers to Dr. Schwartz

Janet's Reading

Subj: **Re: A SPECIFIC PERSON FROM SUSY . . . .**

Date:11/5/2002 5:05:54 PM US Mountain Standard Time

Hi Gary,

I have asked Susy to bring in (as you requested—John) Before we begin I need to say that I now understand why I was hearing the song Tears in Heaven from Eric C. I believe not only was this to show me that it was a "son" connection but also possibly this could be to show me that the father made some kind of tribute to the son thru music? I can't be sure here since I don't know this person however the music seems to be a connection. It would be great if the music was a big part here. Seems so to me! Besides I have been hearing music in my head all day! As if someone was setting the stage.

ALTHOUGH I'M SURE ERIC CLAPTON'S SONG WOULD RESONATE FOR ANY ONE WHO LOST A CHILD, I DEFINITELY LISTEN TO IT WHEN IT IS BEING PLAYED AND I ENJOY IT. MORE SIGNIFICANTLY, JOHN WAS IN THE JAZZ BAND IN HIGH SCHOOL (THEY PLAYED IN THE THREE MAIN JAZZ FESTIVALS IN EUROPE ONE

SUMMER). A MUSIC SCHOLARSHIP WAS SET UP BY A FRIEND OF OURS IN HIS HONOR AT THE SCHOOL, AND IS GIVEN AT THE END OF EACH SCHOOL YEAR. SOMETHING ELSE COULD BE THAT ONE OF HIS FRIENDS MADE US A TAPE OF HIS FAVORITE SONGS AFTER HE PASSED.

OK, now we begin again in the same way . . . . I see a school . . . . actually I am being shown my son's grade school: St. Francis of Assisi . . . . I see the playground and the kids voices in the background. I have the impression that this son had passed young. Maybe because I am only shown a grade school? Or does it have to do with animals? Since St. Francis was the patron Saint of the Animals? Hm . . . .

JOHN'S BROTHER WENT TO ST FRANCIS HIGH SCHOOL, BUT NOT JOHN. JOHN WAS TWENTY TWO WHEN HE PASSED.

OK, now he comes in and not so clear . . . I see grays around the chest area and head as well . . . . To me this shows some kind of brain and maybe heart or lung connection. Could it be immune system or something connecting the two . . . . brain damage or relating to the brain and heart . . . . Although I hear laughter . . . . rather giggles . . . . I'm not sure why but I do hear this.

I also am seeing the "Woody" doll from Toy Story for some reason . . . . Kind of being flipped around and the whole Disney theme . . . . To me the first movie stands out. Whether it's the movie itself or the theme I'm not sure . . . .

JOHN DIED FROM BLUNT FORCE TRAUMA TO THE HEAD AND NECK AND WAS THROWN FROM A MOTORCYCLE, I.E., "FLIPPED AROUND". JOHN'S

INJURIES WERE TO THE HEAD AND CHEST AND HIS BROTHER RICKY'S DEFECT WAS THE HEART AND LUNGS, SO IT COULD POSSIBLY BE THAT THEY WERE TOGETHER HERE, HENCE THE GIGGLES, ETC.

I also have the idea being sent that there is a girl? Could this be a sister? I think if she has lighter hair then it is . . . . He says she sees him? Or has seen him . . . . Some young girl has seen him. And he shows some kind of starry light? or night light with stars? Something that lights up with stars.

AMANDA!! AMANDA WAS KILLED NOT LONG AFTER JOHN IN A FREAK ACCIDENT WHILE LYING ON TOP OF A MOVING SUV AND LOOKING AT THE STARS ON A BEAUTIFUL SPRING EVENING. SHE IS THE DAUGHTER OF FRIENDS OF OURS, WAS ANGIE'S COLLEGE ROOMMATE, AND JOHN KNEW HER.

I'm not sure how this fits but I see jeans and a belt . . . . Something relating to jeans and a certain belt? Hm . . . .

Did John have a problem with mushy foods? Because I hear something about not liking mushy foods . . . . (and laughter). He was more the crunchy kid . . . . liked to chew. Actually I get the chew, chew, chew feeling . . . I know that is odd to say but it is what comes thru . . . .

Oh and about the jeans and the belt . . . . Would there be a relation to a cowboy? I see something with this . . . . Unless it all relates to Toy Story? Hm . . . .

HE WORE JEANS AND BELT MOST OF THE TIME AND HAD COWBOY BOOTS.

WE DON'T KNOW ABOUT THE MUSHY FOODS BECAUSE HE ATE SOFT FOODS.

THE ONLY THING THAT COMES TO MIND REGARDING CHEWING IS THAT AS A KID HE AND HIS BUDDIES CHEWED TOBACCO, BUT IT WAS NOT A HABIT.

OK and he flips back to the music . . . . Who plays the piano? He shows he is banging on the keys . . . . to make his point, not that whoever plays does this.

DAD (ME) TRIES TO PLAY THE PIANO, AND BANGED THE KEYS TOO HARD WHEN I USED TO PRACTICE. JOHN HAD COMMENTED TO ME ABOUT THIS ONE DAY.

OK and the rainbow is over the Mothers head . . . . However there IS a MOTHER figure over there and she is something else! haha (She comes across like right out of a Hollywood movie type role, dressed with class!) I feel like there is a black and white picture of her as if to say . . . . Look here dear! She's a trip!

Yet the mother here must have something with colors . . . . waves of colors and he shows the rainbow over her head. Just streamers of colors . . . . and a bell? Hm

THIS COULD BE DIANE'S MOM WHO WAS INVOLVED IN LOCAL PLAYS. WHEN SHE WAS OLDER, SHE DID DYE HER HAIR DIFFERENT COLORS AND SOMETIMES WORE WIGS. SHE ALSO HAD A BELL COLLECTION.

He doesn't go past the grade school ages with me . . . . so I feel he was young when he passed. He leaves me with his arms wide open . . . . whether to give a hug or to encompass more I don't know.

HE PASSED AT 22, WHICH I CONSIDER YOUNG.

I can say from what I received he's happy and he has friends. Like I said before . . . . with others that have passed like he did.

HE WAS ALWAYS A CHEERFUL KID, RIGHT FROM THE BEGINNING. THERE WERE SEVERAL KIDS FROM PORTOLA VALLEY AROUND THE SAME AGE THAT ALL PASSED WITHIN A FEW YEARS OF EACH OTHER.

I apologize if anything comes across too strong . . . . I just go with how I feel, what I see and what he/they project.

He does say something about his face/head being OK? Not sure how to take that . . . . As he drifts back it almost looks like he was riding a . . . . haha . . . cow? No, that couldn't be, could it? Oh wait, maybe it was a SPOTTED pony? Hahaha

A MOTORCYCLE.

Well, that's all I have for now.

I wish you Gary and the parents Blessings and peace!

Janet . . . . Maybe the song "Tears in Heaven" has a special meaning . . . it just stands out so strong. And he keeps bringing me back to music for some reason. Hm . . . . Well . . . .

End of Reading.

I don't know what your first reaction to this is, but after I thought about it for a short time, I was pleasantly surprised and amazed. In sending the request to Janet, Dr. Schwartz gave her only the first name of our son. How many people named John are there in the world? And yet, Janet came up with several things that applied to our situation.

1. Music was a major part of his short life. Most of us listen to music and might consider it an important part of our life, but the fact that he was in a band that played in major festivals makes it a significant hit.

2. The father made some kind of tribute to the son thru music. (This was made possible through the generosity of a friend). A music scholarship in memory of someone is definitely a tribute, and adds to the credibility of the statement above. Solid hit.

3. She states: "OK, now he comes in and not so clear—I see grays around the chest area and head—", and follows this with references to being "flipped around." The fact is that John suffered blunt force trauma to the head and chest area. He had been thrown from a motorcycle and propelled into a metal, "no parking" sign pole. If her statements had indicated that he died from colon cancer or heart disease or drowning or a host of other possibilities, it would be a miss. But the images she reported are consistent with the way he died. Solid hit.

4. A reference to a young girl being with our son followed by comments about a starry night. As I explained in Ch 5, Amanda was a young girl that grew up in our town and went to both high school and college with our daughter Angie. She was killed in a freak accident one spring night while lying on her back on the roof of a moving SUV with a young man and looking at

the moon and stars. When I read that, my skepticism suffered a huge blow. Another solid hit.

5. Dad banging on the piano. Another hit.

The other comments, though correct, do not get rated as high. For example "I feel he was young when he passed." He was certainly young, but because of the comments about the grade school, it scores slightly lower. The comments that he wore jeans and a belt, about a mother figure who was in plays, about the black and white picture, etc. are all true. The comment about riding a cow? It gets rated as a miss, and yet it's the kind of humorous comment that John would make in reference to riding his motorcycle—more like riding a bronco—and the accident did happen in farm country.

Another point: If we accept for the moment that the images Janet was receiving were from John, then this supports the concept that the deceased know what is going on in the physical world. Why do I say that? Because the movie *Toy Story* came out in 1995, four years after John died.

Janet had no idea when doing this reading who it was for, other than that the deceased was someone named John. She didn't know the age of the deceased or anything about the circumstances surrounding his death. The comment about the Woody doll being flipped around could have had a number of different meanings: perhaps the sitter had gone to that movie with the deceased, or perhaps it was a toy given to a child, or any number of other possibilities. When Janet found out about the circumstances that prompted that image (as interpreted by me) she was concerned that her presenting it the way she did made her appear heartless. I assured her that this was not the case. She was simply reporting an image that she saw. However, I think this points out how difficult being a medium can be, especially one as compassionate as Janet. Mediums are given images, sounds and/or feelings and can only report them for the sitter to interpret. It should not be forgotten that they are very human, everyday people who also have families, and these events can have a big impact on their emotions and feelings.

After reviewing this and other readings, I began to get the picture that the discarnates are trying to come up with as many things as they can to let us know it is actually them. How many people would the above *combination* of specific comments apply to? In the next chapter I present another reading and show how it may be, for the discarnates, like playing a game of charades.

The John Kaspari Fund

After commenting and scoring all of the readings, I forwarded them to Dr. Schwartz. We discussed them and I began to get a better understanding of the basics of mediumship research. Over the next couple of months we communicated about how we might be able to work together, and discussed possible experiments. On March 5, 2003 I made a commitment to establish the John Kaspari Fund at The University of Arizona. It was now time to decide what our first experiment should be and how it would be conducted. Because of the readings I received I had become convinced that there really does seem to be an afterlife, and I was developing a strong desire to know what it was like. Questions such as—What do you "do" there? Was it similar to life on Earth in that there were ethnic groups? Were there religious groups similar to (or the same as) those on Earth? Were there males and females? Did people who committed serious crimes on earth such as murder, child molestation, etc. receive punishment in the afterlife?—and many other questions came to mind. Dr. Schwartz, Dr. Julie Beischel (a researcher who worked with Dr. Schwartz at the time) and I met at my hotel in Tucson and decided that the name of our experiment would be "The Asking Questions" experiment.

Since the discarnates seemed to be willing to provide information, even though it may be only to let us know they still exist, we were going to see if the mediums could ask questions and get verifiable, meaningful answers.

Around this time another couple, Bob and Phran Ginsberg, who had recently lost their daughter Bailey in a tragic accident, had read Gary's *Afterlife Experiments* book and had also contacted him. Gary, Julie and I decided that we would do a pilot reading first to try to determine if mediums can even ask questions of the discarnates. Diane and I and Bob and Phran would be the sitters. These experiments were conducted as a part of the Veritas Research Program at The University of Arizona. (This program closed in 2008).

Although they were preliminary investigations, the experiments produced very encouraging results, and indicated that mediums can indeed ask questions of discarnates and receive answers. As I stated previously, I do not intend for this to be a scientifically rigorous book. Rather than go into the details of our experiments, in the next chapters I am going to try to give the reader a better understanding of mediumship. In Chapter 16 I introduce two recently published books whose authors have investigated more than 150 years of data, as revealed through mediums, that suggest answers to some of the questions we were asking in our experiments. Neither of these books were published when we did our studies, but they offer an image of the afterlife in far more detail than I imagined possible when I began my search.

One of the most significant accomplishments of the Veritas Program was that the experiments were conducted following increasingly rigorous scientific protocols. A great deal of time was spent working on the test and scoring protocols to be sure they met accepted scientific standards. I discuss briefly a couple of these criteria at the end of Chapter 7, but a much more detailed description is presented in papers by Drs. Beischel and Schwartz. For those who have an interest in the science behind these studies, I reference websites in the back of this book where these papers can be reviewed. This may not be of interest to the general public, but doing studies this way is necessary to achieve credibility within the scientific community. One of my goals in writing this book is to provide information that will help skeptical scientists open their minds to the concept of an afterlife. My hope is that if this goal is achieved, it will become a part of mainstream science and as a result will serve to help those without a belief system who are suddenly faced with a devastating loss. Perhaps we can begin to move the pendulum back toward the center.

Prior to my involvement with Dr. Schwartz, he had been conducting scientific studies of mediums that resulted in his *Afterlife Experiments* book, and these studies continued after I became involved. As I mentioned, one of the three mediums who emailed readings for me the night after I met Dr. Schwartz was Allison DuBois. In January 2005 NBC premiered a show titled *Medium* that was inspired by Allison's life. This show ended on CBS in 2011. Dr. Schwartz had been testing Allison's mediumship capabilities prior to the debut of that show, and in 2005 he published a book entitled *The Truth about Medium*. One of his reasons was to make the general public aware that there is more to the subject of mediumship than the impressions given by typical "Hollywood" movies, that in fact this was a phenomenon that was exceeding expectations in scientific tests conducted in a university laboratory environment. Dr. Schwartz' book chronicles many of the experiments he had been conducting.

One of the problems when attempting to do research outside of the mainstream areas is that very little funding is available. As much as I wish I had been able to continue our project, I was not able to contribute additional funding. Also, because much of the funding that universities receive is from the government or from people with other interests,

Dr. Schwartz had to devote more time to work funded by the National Institute of Health (NIH) in areas they felt were more important than something as far out as afterlife research, so our "Asking Questions" project was put on hold. However, I had gained enough insight into the subject of mediumship that I was now able to examine it with an open mind, which in turn opened my mind to the general topic of "paranormal" phenomena.

CHAPTER 7

Playing Charades

Do not go where the path may lead; go instead where there is no path and leave a trail.

Ralph Waldo Emerson

Soon after becoming involved with Dr. Schwartz I learned of a new website called TestingMediums (TMs) that had been started by a few people with similar interests to ours. Originally, the general purpose of the website was to promote discussion about mediumship, and how best to test mediums, but it has since become a more general paranormal phenomena discussion site. It was put together by Steve Grenard, a doctor in New York who had also lost a son. There were initially several people involved in the discussions, mostly from the United States and Great Britain.

One of these people was Peter Hayes, like me an electrical engineer. Peter was a PhD, had very rigid scientific principles, and had committed to provide funding to Dr. Schwartz for a fellowship endowment so that he would have someone to help in his lab. Soon after, Dr. Schwartz met and hired Julie Beischel, PhD as his William James Fellowship recipient. (Julie turned out to be an outstanding researcher.

She and her husband Mark Boccuzzi have since started their own research organization called "The Windbridge Institute").

During one of the TMs online sessions, we were discussing the pros and cons of various scoring methods for future readings, and I had submitted part of a reading from Janet Mayer as an example for anyone interested in attempting to score it. This was the first reading that Dr. Schwartz had emailed to me, and as I stated earlier, when I first started going through the original readings, I was a little confused. I had expected it to be as if she were there talking directly to the deceased, as you sometimes see in a movie about "psychics" and the afterlife. After I learned that the mediums are experiencing mental and physical sensations and then must try to interpret what they are sensing, it occurred to me that, for the discarnates, it could be like playing a game of charades—you must present clues in the form of "sensations" to a medium and see if they can get the message you are trying to send.

One suggestion being proposed at the time for scoring the readings was that each statement should be evaluated as a stand-alone statement, and not taken in the context of the entire reading. I did not think this was a good idea. To me it was like having a discussion, or listening to a lecture, where you were not allowed to evaluate each new sentence in the context presented by the entire discussion or lecture. In order to explain my view, I used the information presented by Janet in this reading to make my point about the concept of charades. I am first going to present a part of the reading that I had submitted for scoring, and will then show the letter I posted on TestingMediums to explain my position. (I explain at the end of this chapter how we improved the scoring by adding additional criteria and also how we began doing multiple blind studies).

Below is Janet's reading. It is a long reading, so I broke it into sections. Following the first section, I show how it can be compared to playing charades. When I received this reading I was expecting to hear from my son John, but it turned out that, in the initial part of the reading, the statements presented fit my father rather than my son. One of the things that convinced me, in addition to most of the information being valid for him, is that the personality of the deceased also fit my father, who was a man with a very strong personality. If you knew my dad, you would understand why it is a long reading. The next section will show how Diane's father came through following my dad.

Parts of the reading may seem confusing, as they did to me at first, but keep in mind that the mediums have no control over the images, sounds or feelings they are receiving. All they can do is report them. Dr. Schwartz would ask the mediums to "just report what you get—don't try to interpret it. Let the sitter do that." To do so is apparently easier said than done.

Section I of Janet's Reading.

Subj: **Susy Reading . . .**

Date:11/4/2002 8:21:22 AM US Mountain Standard Time

Good Morning Gary,

Before I begin I need to clarify something. You asked that "we" do a special reading to bring in "his" loved ones.

Below this you state you have not told the sex of the person you met. However you did . . . . You said it was a he. I just thought I should make note of this.

COMMENTS IN ALL CAPS BY BILL & DIANE KASPARI

I actually have two men that Susy shows to you standing near . . . .

She says one man lost his father not too long ago and one lost his mother . . . . a while back.

UP TO THIS POINT, DOESN'T SEEM TO FIT, SINCE THE MEN ARE THERE.

I think this is for two different people. Unless this is the son's father who lost his wife then the son lost his father? That seems a strange way for Susy to bring this thru. Unless it fits two different people . . . . We'll see.

I AM THE SON, MY FATHER LOST HIS WIFE MANY YEARS AGO, THEN I LOST MY FATHER 12 YEARS AGO. THIS WOULD BE MY FATHER.

OK, now I see what to me looks like farm animals. However, they don't seem to fit. When I say this it's because I see a pig, then a goat or well it might be a sheep and then I see a snake and a dog! They don't really blend in with a farm setting but this is how Susy shows them to me!

WE LIVE IN A RURAL AREA, BUT NOT FARM COUNTRY, AND THERE ARE A VARIETY OF ANIMALS IN THE AREA.

I also see lots of traveling . . . . Susy shows me a schedule or calendar. She shows traveling to different countries? . . . . Hm . . . . She keeps taking me out of the country and I see foreign lands. Ok, this man I see that comes in . . . . I almost think he is talking in a foreign language! haha

MY FATHER TRAVELLED AND WORKED IN MANY FOREIGN COUNTRIES FOR MANY YEARS, BUT HE SPOKE ONLY ENGLISH.

That can't be, can it? It's as if I hear something but feel sounds from him. He keeps showing me his throat? Something about the voice . . . .

MY DAD HAD A THROAT DISORDER WHEN HE WAS OLDER (HIS THROAT WOULD CLOSE UP AND HE HAD TO PUT A MERCURY FILLED LEATHER TUBE DOWN DAILY TO HELP KEEP IT OPEN),

He is also showing me poverty . . . . or that when he was growing up there was poverty around him but not in his mind. I hear what sounds like "work hard, believe, it will happen" . . . . have faith . . . . yes . . . .

THIS FITS MY DAD VERY WELL. HE WAS POOR AS A CHILD, HIS DAD DIED WHEN HE WAS SIX AND HE HELPED HIS MOTHER TAKE IN WASH, ETC. THEY LIVED IN A RAILROAD BOXCAR FOR A WHILE. HE WORKED HARD, INCLUDING WORKING HIS WAY THROUGH COLLEGE, AND TAUGHT US TO DO THE SAME. PERHAPS THE "BELIEVE" COMMENT REFERS TO THIS ENDEAVOR.

This man must have been very Spiritual for that is the feel he projects to me.

MY DAD WAS VERY RELIGIOUS, WENT TO CHURCH REGULARLY AND WAS ACTIVE IN HIS CHURCH.

Now this man shows me flowers that I relate to Hawaii. As if the garden is very important here. Or the scene of the garden?

MY DAD RETIRED IN HAWAII AND WAS ALWAYS WORKING IN HIS GARDEN. THE SCENE OF THE GARDEN COULD HAVE TWO MEANINGS: (1) IT WAS THE SITE WHERE MY DAD HAD HIS HEART ATTACK. (2) MY SON JOHN AND I VISITED HIM NOT LONG BEFORE MY FATHER'S (AND JOHN'S) DEATH, AND WE DID SOME TREE TRIMMING AND YARDWORK FOR HIM.

Oh and the flowers are reds . . . . bright reds!

THE HOUSE IN HAWAII HAD BOUGAINVILLE GROWING ON A TRELLIS IN FRONT OF THE HOUSE, AND MY SON JOHN AND I TRIMMED THEM FOR HIM. I DON'T REMEMBER FOR SURE, BUT THEY WERE PROBABLY RED.

Now I see two men below this man so I guess this would be the son, and grandson he speaks of now?

NONE OF MY DAD'S SONS HAVE DIED YET, AND THIS MAN CERTAINLY FITS THE DESCRIPTION OF MY DAD. HOWEVER, I HAVE LOST TWO SONS, SO PERHAPS THEY ARE THE TWO MEN (I.E. BOTH GRANDSONS).

He talks of Hactic? or something that sounds like this? Hm . . . .

THIS ONE INTRIGUES ME, BUT NO CLUES YET.

This "father" figure is so proud of his son . . . . it's as if he has oceans of energy just washing over and I feel such a sense of peace here. Water comes in very strong . . . . cleansing and healing.

PEOPLE OFTEN COMMENTED TO MY BROTHERS AND I ABOUT HOW PROUD MY DAD WAS OF HIS KIDS. I CAN ONLY SPECULATE ABOUT THE WATER (HAWAII, I WAS IN THE NAVY, ETC.), OR PERHAPS IT'S JUST TO CONVEY THE FEELING OF CLEANSING AND HEALING.

Gary, I don't think I have ever had such a reading like this . . . . I just feel such spiritual energy coming from this man that I see nothing besides him and what he shows . . . . there could be 10 deceased relatives but this man just stands here.

He has dark hair, very dark . . . . and wears a ring . . . seems very gold and clunky.

(He held his hand out to show this)

MY DAD HAD GREY HAIR AS FAR BACK AS I CAN REMEMBER, AND IT WAS PROBABLY NOT VERY DARK WHEN HE WAS YOUNG. HE DID WEAR WHAT I BELIEVE WAS HIS COLLEGE GRADUATION RING.

For some reason he tells me he is barefooted? Not sure why, maybe it has some importance?

I DON'T KNOW WHAT THIS WOULD MEAN. HE WAS NOT REALLY A BEACHY TYPE PERSON. PERHAPS IT IS THE POVERTY BACKGROUND, OR HAWAII CONNECTION.

I have to ask but was this man a Spiritual leader of some kind? He shows me the third eye and he shows a woman I know that wore a jewel on her third eye and the wrap of material around her. (I can't tell if he was saying she was eccentric or from India?) Hm . . . .

MY DAD WAS RELIGIOUS, AND BELONGED TO AND WAS ACTIVE IN THE KNIGHTS OF COLUMBUS. REGARDING THE WOMAN, THE CLOSEST I CAN GUESS WOULD BE MY MOTHER, WHO WAS OF ITALIAN DESCENT AND WAS BORN IN ITALY. REGARDING JEWELRY AND CLOTHES, I CAN'T SAY BECAUSE SHE DIED WHEN I WAS AN INFANT.

End of Section I.

There is a lot of information in this reading. A very significant portion is solid hits while some comments are difficult to evaluate and others appear to be misses. For example, I didn't know how to interpret, let alone score, a comment such as "a jewel in her third eye" since I had never even heard of such a thing. I had to go to Wikipedia to learn that it is "a mystical and esoteric concept referring in part to the brow chakra in certain eastern spiritual traditions." There are other interpretations also listed there. Again, because of my scientific mindset, I didn't know what to do with that. I couldn't say it was wrong, so it didn't get scored.

There were some comments that were wrong, and they were scored accordingly. However, as I later learned after evaluating these original readings, I sometimes was either missing the connection or just didn't remember something. (I didn't think of this until much later, but the comment about "Hactic" may just be a typing error. Perhaps she meant "Hectic" which would describe my dad's life, but that is too speculative to use for scoring purposes). Think how difficult it must be for the medium to receive the images or other sensations and then try to understand them while recording what they are sensing. In order to help the reader understand the medium's dilemma, Chapter 9 gives a little insight into the process.

As I stated previously, while evaluating the readings I had received, it occurred to me that, for the discarnates, it may be like playing a game of charades. Peter Hayes had attempted to score this reading using the criteria we had been discussing, one of which was to treat each comment by the medium as a stand-alone statement. One of his observations was that we may lose the ability to get the full sense of the reading if we do this. I was in strong agreement with his observation. Below is an email I sent to the TestingMediums group in response to Peter's comments. Keep in mind that it was written primarily for scientists, so I apologize in advance to those who are not interested in the scientific aspects.

My "Charades" interpretation of this reading.

A reply to Peter's posting: *Some remarks on A SAMPLE READING.*

> I think that Peter has already gotten to the core of the scoring problem. I would like to make what I believe is an important point. Near the end of Peter's DISCUSSION, when he talks about the difficulties he encountered with the present scoring system, he comments that "some of this may be due to total context, which is lost in breaking it down into segments". I believe this is definitely true and a very important point. When I attempted my first scoring (accuracy only), I tried my best to be as objective as possible, and as previously stated, I felt additional factors were needed.
>
> However, after a few subsequent scoring attempts and continued uncertainty, I came to believe that in trying to score the statements with pure "scientific"

objectivity even with the additional attributes, I was not allowing myself to "***look at*** the forest for the trees", although the forest was becoming visible. I began to realize that my supposed objectivity was biasing me against the medium. By looking at each statement without putting it in the context being presented by both preceding and following statements, I felt I was *losing* objectivity.

As I later commented, a factor that I became aware of was the *personality* of the individuals coming through which gave me a sense that the readings were authentic. I then asked myself if I was guilty of doing what debunkers might say—something like "Oh sure, he lost his son and wants to believe in this stuff anyway". While I understood that this was possible, I didn't see how I could be functioning normally in every other aspect of my daily life, but not where my objectivity was concerned.

I began to think about what it would be like to be on the "other side" where it was up to me to in some way get information across to a medium, assuming that I was essentially the same person that I am now, only in spirit "form". I concluded that the first thing that must be done would be for me to identify myself by giving data to the sitter that is accurate, meaningful and (to use Peter's new term) rare. Could I do it with discrete sentences? I concluded it would be quite difficult unless I were able to have an actual conversation. It then occurred to me that the process might be compared to playing Charades—you must get the message across but you are not allowed (or able) to talk.

After doing this soul searching, I decided to go back and score the readings trying not to be overly scientific. I can only say that I was much more comfortable analyzing and scoring the readings.

Now let us examine the information presented.

If we look at the statements that Peter has scored so far, **what are the facts?** I am going to paraphrase some of the statements and/or my comments to save time, and put them in quotation marks for identification purposes.

1. "I lost my father not too long ago and my mother many years ago." My father died in 1990, which some people would not think of as "not too long ago". However, my mother died in 1937, so now the statements "not too long ago and many years ago" make more sense, even though they are not scientifically precise terms. In general, the statements fit my situation. However, because of the uncertainty as to whether there are one or two people, it is difficult to rate this segment.

2. "Now I see what to me looks like farm animals, but not in a farm setting." The basic facts are that I do live in a rural area, but not farm country, and the animals mentioned are in the area. This is somewhat vague, so I would give it moderate scores. However, we are still in what I will call the "introductory" phase of this reading.

3. "I also see lots of traveling. Susy shows traveling to different countries. I almost think he is talking in a foreign language." The fact is that my dad did work and travel all over the world for many years. I would say that the comment about the foreign language was the mediums attempt to interpret what she was seeing.

4. "That can't be, can it? It's as if I hear something but feel sounds from him. He keeps showing me his throat? Something about the voice—" It was about here that what skepticism I still had began to dissipate. How many people would be showing something about their throat? The fact that he had to put this long, mercury filled tube down his throat every day made this, in my opinion, an accurate, very rare and highly meaningful statement. I would view the comment about the voice as the medium's interpretation of the image she received.

5. "He is showing me poverty (when he was growing up)." The fact is that my dad's father died when he was six, for a while his family lived in a railroad boxcar in Kansas, his mother took in washing, etc. What made this segment more believable was that he used to talk about it occasionally, so he could expect that it would resonate with me.

6. "This man must have been very spiritual." The fact is that he was very spiritual, and he would expect me to consider this a significant fact. He was buried in his Knights of Columbus uniform (this has significance in another reading by another medium).

7. "Now this man shows me flowers that I relate to Hawaii. As if the garden is very important here. Or the scene of the garden?" These statements are dead on. He retired in Hawaii and worked in his flower garden all the time. The last time I saw him, only my son John was with me, we all worked in the garden, and it was where my dad had his heart attack.

Playing Charades

So to summarize all the facts in the first seven statements, an image was created for me of a spiritual man who grew up in poverty, was a world traveler, retired in Hawaii, spent time working in and had significant events occur in his flower garden there, and had a problem with his throat. In addition, he pointed out that he died "not too long ago", my mother died many years ago and I lived in a rural area. Looking at this from the "charades" viewpoint, I felt he did a pretty good job! After all, he had never done this before. How many people would this collection of facts apply to?

Further on in this and other readings, my dad's very strong personality becomes more obvious. I think the fact that he was the first person to come through demonstrates this. I will also note that there appear to be some significant misses later. While these are most likely just misses, it could also be that I may not be getting the message, or may not be remembering something. I talked to my wife and brothers about a couple of things while trying to do the scoring, and we sometimes had different recollections.

I will list some major points as I see them.

1. As stated by others previously, the quality of the readings depends on the discarnate's ability to play "charades", the mediums ability to receive and interpret the images or other symbols, sounds, etc. received and the sitter's ability to interpret and make comments about the statements.

2. As stated previously also, we should not expect mediums to be scientists. It should be up to us to analyze their interpretations of what they receive when appropriate.

3. We can not expect to score readings by looking only at discrete statements. We must look at the entire reading, or sections of a reading, in context. (I use the word "section" to define, e.g. a group of segments relating to a particular individual or situation).

4. Only the sitter(s) can be expected to interpret the statements, and probably some sort of "training" or help will be required.

5. We can not expect most sitters to recognize or remember all of the facts immediately.

I believe the scoring system should somehow incorporate these points. I don't know what the Robertson/Roy system consists of, but if it helps address some of these points, I would like to review it.

Comments are really appreciated.

Best Regards,
Bill Kaspari
March 20, 2003

Section II

This section is a continuation of the reading where I left off in Section I. Janet is suddenly getting images from a different person, as evidenced by her comments.

This one man isn't bringing anyone else in for me . . . . he wants me to know he is HERE!

I also see smoke around him . . . . as if he is sitting down and smoke rises around him.

MY DAD WAS A HEAVY SMOKER (CAMELS BEFORE THE DAYS OF FILTERS) AND SMOKED THEM UNTIL THERE WAS NOTHING LEFT TO HOLD. (HE NEVER WASTED ANYTHING). HE QUIT IN HIS MID-FIFTYS AT DOCTORS URGING.

HOWEVER, DIANE'S DAD WAS A CHAIN SMOKER, AND SMOKED FOR 56 YEARS BEFORE BEING DIAGNOSED WITH LUNG CANCER.

He is also showing me some kind of spot on his head? Like a sun spot? or black spot? Hm . . .

DIANE'S DAD HAD CANCER SPOTS REMOVED FROM HIS BALD HEAD. HE ALSO FELL ON HIS HEAD A FEW TIMES IN LATER YEARS AND HAD MINOR HEAD INJURIES.

I'm not sure where you are Gary and who you met but I'm feeling warm climate coming from this man on the other side.

GARY WAS IN CALIFORNIA. DIANE'S DAD WAS A NATIVE OF CALIFORNIA.

He is projecting peace . . . . and he seems to have a number of "items" with him?

Items from loved ones? Beads . . . . something that looks like a scarf . . . .

DIANE'S DAD ALWAYS WANTED THINGS TO BE PEACEFUL. BEADS COULD BE ROSARY BEADS FROM HIS MOTHER WHICH HE GAVE TO DIANE. THE SCARF IS MORE VAGUE.

And something he holds up that, well, it looks like a comb? But is this a joke? As if I didn't need this or

I ALWAYS needed it? Something about a comb . . . . hm . . . .

DIANE'S DAD WAS BALD AND HER FRIENDS USED TO CALL HIM THE "BALD EAGLE".

OK, now Gary as you can see I had TWO completely different groups or families come in . . . . which came as a surprise to me since they didn't seem to fit.

Yet, who am I to say? I just send what I get!

Hope Susy knew what she was doing here! hahaha

Many Blessings!

Janet

There was more to this reading, mostly what was apparently Diane's mother coming through following her dad, but I think the above is sufficient to demonstrate the main points I want to make, which are:

1. For the deceased, trying to pass information to the sitter through a medium may be like playing a game of charades.

2. It shows that you do not always hear from the person you are expecting to hear from. Although I was expecting (and hoping) to hear from my son John, my dad came through. What this indicates to me is that things on the other side may not be that different from here. My dad was always the one to take charge or go to the head of the line.

3. In addition to my dad, Diane's parents were there.

4. Janet comments that there were a number of other people there. I think this paints a good picture of what it may be like on the other side. They somehow knew that we were involved with a medium who was attempting to contact my son. Maybe Susy brought them together.

One observation I've made is that sometimes my readings would start off really strong, i.e., with many solid hits, and then sort of taper off in this sense. It makes me wonder if it just becomes more difficult for the discarnates to come up with additional validating information—as if they run out of ideas. Also, just as in charades, some people may be much better at playing the game than others.

Our New Scoring Criteria

One thing that bothered me about the scoring system in Dr. Schwartz' *Afterlife Experiments* book was that a statement by a medium could be completely true, and yet not be of any real significance. For example, the medium could say that the discarnate had brown eyes, but so do billions of other people. Or a statement can be true, such as "He says to tell you he loves you" but it is not highly specific to that particular person. So I suggested to Dr. Schwartz that, for future experiments, in addition to Accuracy of the statement we add two more criteria in the scoring: Specificity and Meaningfulness. He agreed, so later scoring involved using these additional criteria. As I stated in the Charades article, I felt that we were able to be more objective when using the new criteria. (The definition of these criteria is shown in Appendix A). We later changed Specificity to Rareness as mentioned in the *Charades interpretation* above.

The other criteria added is the implementation of double and triple blind studies, developed by Drs. Beischel, Hayes and Schwartz. These were done to insure that the information being obtained by the mediums was not coming from the sitters or the people conducting the research. My original readings were single blind because the medium had no idea who the sitter was, and was given only the first name of the deceased. In a double blind experiment, in addition to the medium not knowing who the sitter is, multiple sitters are used resulting in a number of readings. The sitters score all of the readings and are not told which reading is theirs until after they have completed the scoring. In a triple blind experiment, an additional lab assistant is employed so that the experimenter who works with the mediums is kept blind to the identities of the sitters and the deceased.

Even with these additional controls, the accuracy was much better than would be statistically predicted.

Conducting experiments in this way, while necessary to gain acceptance within the scientific community, is not without difficulty for the experimenters. In addition, they require a great deal of time and therefore money. For those who are interested, a detailed explanation of how these protocols were developed is presented by Dr. Schwartz in *The Truth about Medium*.

CHAPTER 8

Three Special Messages

Experience is not what happens to you; it's what you do with what happens.

Aldous Huxley

There were other readings too numerous to include here, both from the original nine readings and those done as part of the "Asking Questions" experiments. However, I want to tell about three more examples of mediumship to give the reader a better understanding of what they are like. I am not going to reprint entire readings or sections of readings as I did with the previous ones by Janet, but will only tell about a message received in each of them. The first describes an event that occurred related to a reading by Allison DuBois.

Below is an email I sent to Dr. Schwartz describing what happened.

A Predictive Reading

Jan 5, 2003

Hi Gary,

I had a very pleasant and relaxing holiday season, and I hope you did also.

I have something interesting to tell you about.

We set up a Christmas tree a couple of weeks before Christmas. Diane had bought some new lights for the tree that came with a few special bulbs that, if put in the string of lights, would cause all of the tree lights to blink. We tried the blinking lights, but decided not to use them.

The next day, I was sitting in my favorite spot on the couch and I noticed a single blinking red light on the tree. I knew that I had removed the blinking lights, but even if I had missed one an entire string of lights should have been blinking, not just a single bulb. I began to get up to investigate, but when I did the light went away. When I sat down, it appeared again and continued blinking at a fairly slow rate.

I was really curious, so I "marked" in my mind the spot on the tree where the blinking light was and went to investigate, but no light bulb was there! I went back to where I was on the couch, and it was blinking again at the same spot!! The curious scientist in me had to figure out what was going on, and finally I figured it out.

One of the ornaments was rotating back and forth a few degrees and was reflecting the light from a red bulb to another spot on the tree, giving the appearance of a blinking red light at that spot. When I went to the tree, my viewing angle was different and therefore I could not see the reflected light.

When I first noticed the blinking light, I immediately thought about the comment in Allison's reading about John dated 11/8/02:

"He says there will be a moment around X-Mas that everyone will know he's there. He shows a hanging ornament and says to watch for it."

All I can say is "WOW!"

I know that if the same thing had happened in previous years, I would have been content to figure out what was happening. As I thought about it, the skeptic in me said that it was simply an ornament being moved by the air from a heater vent, so what is the big deal? The answer came back immediately that, even though I was able to explain what was happening, it had been predicted weeks in advance, and was quite specific.

The thing that gives it additional significance is that the specific ornament that was moving is one that Diane bought when we were in Alaska in 1999. John went to Alaska with some friends in the spring of 1989 as an adventure and worked in a cannery there that summer. Diane and I wanted to go to Alaska not only because we had never been there, but also because it was a special place to John.

It is also interesting to note that John referred to "dad's special chair" in the same reading.

"He says he sits in dad's special chair. He's making a joke like there was a time when I wasn't allowed but now I do it all the time, HA!HA!"

I didn't have a special chair, but I did and still do have a favorite spot on the couch, and I was sitting there when I noticed the blinking light.

I think many "scientists" would explain away the above occurrence as a coincidence without giving it further thought. I don't know about you, but I find this a little too unusual to be considered a "coincidence". As you keep saying, it is hard to ignore the data.

I think it points out the importance of keeping an open mind and being as objective as possible if we are to advance our understanding of mediumistic phenomena. Just because we are able to explain some unusual phenomena does not mean that they have no special significance outside of current scientific thinking.

Let me know when it would be a good time for us to meet in Tucson. I know there are others such as Julie to be considered also.

Best Regards,
Bill Kaspari

One of the things that made this occurrence a pleasant surprise for me was that I had completely forgotten about the prediction in that reading, and in fact I remember being somewhat skeptical when I first read it because it seemed to be a rather vague prediction. This was one of the first nine readings that I received, and I still had a high degree of skepticism to go with my lack of understanding of how mediumship worked.

The next message is not taken from a reading, but occurred when going to a lecture given by a medium.

A Surprise Visit from John

On April 16, 2004, Diane and I went to the East-West bookstore in Mountain View, CA to attend a lecture being given by a medium from Scotland named Gordon Smith. His new book on mediumship, entitled *Spirit Messenger*, had recently been published so he was there for a book signing as well as the lecture. He made it clear at the beginning of his talk that he was there only to describe the basics of mediumship—not to give readings. Following his lecture, we were allowed to ask questions as well as get our book signed.

I was curious at that point to know how the deceased seemed to know what we were doing so often, so I asked Gordon: "Are the discarnates around us all the time?" to which he replied "Oh yes, definitely!"—"In fact, there is a young man with you right now—he died suddenly!". Gordon then turned directly to Diane and said "You took something off!" Diane looked at him somewhat puzzled and then he mentioned "a pendant—but not the one you're wearing." I thought Diane was going to faint. She was wearing a black pendant on a cord. While getting ready earlier at home, she started to remove the gold locket necklace which she usually always wears, because the locket made a lump under her "fashion necklace." At the last minute she thought "No, it is especially important to keep it on tonight."

As soon as Gordon mentioned "a different pendant" she brought out the heart locket which was under her sweater. (It was a high necked sweater that you couldn't see through). He nodded and said "That's it. Your son's picture is in it." It was. Gordon went on to say that "he (John) knows there is an anniversary coming up, and that he will be there." (The anniversary of John's death is May 2nd).

I have to admit that I was really impressed, and any remaining skepticism I had regarding mediumship was essentially gone.

I did a book review of *Spirit Messenger* for the *Signs of Life* publication of The Forever Family Foundation, and have reprinted it in Appendix B. It tells a little about Gordon Smith's background.

An Impromptu Reading

This next message also came as a surprise. It was not part of the experimental readings being done at Dr. Schwartz' lab and came after we had completed the "Asking Questions" pilot study. I am not going to reprint the email I received but will relate the key points. In the next chapter I reprint some correspondence with Janet Mayer that gave me a much better insight into what it is like for a medium when receiving the information they report.

One morning in Feb. 2005 when I opened my e-mails there was one from Julie (at Gary's lab) in which she forwarded a short reading from Janet Mayer, who knew it was for me because John (Casper) popped in on her that morning and she felt he apparently wanted to get a message to my wife and me. She recognized John from the previous readings.

There were essentially two messages, the first about a birth and the second about someone who had recently died. As I read through the comments, I was getting no connection, i.e. I was not aware of any births or deaths of close friends or relatives.

Then she says "Also did John's father recently bump his head on the right front side?" That got my attention. No, I had not, but the day before receiving the e-mail when I went to my daughter Becky's house she had her husband show me a huge bump, black eye and cut over his *right* eye. The night before, his pager started ringing after they had gone to bed so he jumped out of bed to turn it off before it woke their daughters, and walked into the corner of a wall in their house! I was amazed, especially since it was something that had just happened.

I then talked to Diane about the comments related to a birth and death. Diane was supposed to attend a baby shower that day for a good friend's son and daughter-in-law. She received a phone call early that morning that the mother-to-be was in the hospital delivering the baby, who was premature, and the shower was cancelled. My son John had been friends with the new dad.

Then I asked Diane about the comment related to someone who recently passed being with John. Diane is a drug and alcohol addictions counselor and had a client whose girlfriend had recently died, and Diane's client was struggling. Diane had been concerned about this client and how best to help him, and says that the night before I received the message she lay in bed and "talked" to the girl who died asking her to help her client and his family. Diane believes that the girl was letting her know that she heard her, and believes it could be the main message of the reading.

This was the last reading I received and by now I was convinced that the information being presented by the mediums had to be coming from deceased people. Some scientists have offered other explanations. One is that mediums are reading the sitters mind, but this could not explain much of the information I had received—for example, the Christmas tree ornament prediction. Another theory is that all information is stored in some sort of universal energy field that is being accessed by the mediums using their special sensitivities. While this is theoretically possible, I find it easier to accept that we simply survive the death of our physical bodies. As I explained previously, one of the reasons is that the personalities of the deceased come through. I suppose it's possible that our personalities are also stored in the universal energy field. I have trouble accepting theories like this because they appear to me to be attempts by scientists to explain afterlife phenomena by any means other than survival of bodily death.

Having been convinced that we do survive bodily death, and that Janet was receiving communications from my son, I finally began to realize that the deceased must be aware of our daily activities. How else could John know about something so specific as my son-in-law getting a bump over his right eye? I didn't yet have any idea how this could be, but I could come up with no other logical explanation. I finally concluded that we may be in a sort of "fishbowl" and can be viewed by the people on the other side whenever they want.

CHAPTER 9

Understanding Mediumship

> *It's not what you look at that matters, it's what you see.*
>
> *Henry David Thoreau*

As I began to see an increasing amount of evidence supporting the validity of mediumship, my doubts were replaced by a curiosity to better understand what mediums actually experience when they are receiving these messages.

I decided to email Janet Mayer in the hope that she would be willing to answer a few questions about the sensations she received during a reading. Janet very graciously agreed to answer my questions. At that time, I had posted some comments on the TestingMediums website telling about the last reading I had received. I called it the "Casper/Bump-on-the-head reading."

The following is a post I submitted to TestingMediums. This posting was initiated by a query from David Haith, one of the TMs participants, who had expressed an interest in the circumstances surrounding the reading. Here is my posting:

"At the end of my post to Dave on Feb. 5, when answering his questions about the circumstances under which the Casper/Bump-on-the-head reading by Janet Mayer was given to me, I said that I hope to find out from Janet what the actual image was that she saw. Was it a man, did he look like my son, my son-in-law, etc.? Remember, she did not state it as a fact but asked if I had struck my head.

I was very pleased to receive the following reply from Janet which I think is very informative. My original intent was to simply explain what she told me, but as I told Janet, I don't think I could explain it nearly as well as she did, so I am forwarding her reply (with permission from Janet and with Gary's blessing). I think it will help us understand how the mediumship process works, which as Peter Hayes pointed out was the main reason this forum was started.

Janet's reply follows, as she originally sent it to me."

Hi Bill,

I'll try to answer your questions as best I can . . . .

First, let me explain that from what I can recall that morning, I believe shortly after I woke up I "felt" someone was around me. So I already had the feeling I was about to find out something from someone on the other side. After my son left for school and I began to start my day I suddenly heard "Casper" and stopped short. That triggered an immediate "person" for me. After all, John is the only person I relate to with that name. So I asked: Is this John? as I walked into the computer room and sat down. I decided to just sit down and type whatever came thru right away so I wouldn't lose the connection.

So, first off, I felt someone around me. I couldn't see the Spirit/person but I felt it. From having this "ability" for so long I knew it was someone wanting to come thru. The next thought was triggered by the word "Casper" that I heard. Not outside my ear this time but inside. Before I move on let me explain this. I can hear Spirit voices at times as if a person is speaking to me; however, I usually hear a name or word in my inner hearing. Does that make sense? So, I unfortunately don't hear whole sentences or paragraphs, just a word here and there. Perfect example is when you're on the cell phone and it cuts out. You may hear: Bob . . . . cleaners . . . . Chinese . . . home. So I would take that as: Hi Bob, I'm running to the cleaners, then I'll grab some Chinese food and head home. It's the same sort of concept for me. I take the pieces to create the puzzle. I try not to interpret, but at times I'm at a loss for how else to convey a message. So, I do the best I can and try to keep it as close as possible to the way I hear it.

Let's say John feels the "need" to get a message across for whatever reason. He is only given a small amount of time (for whatever reason) so he has to give me what he can in as few visions/thoughts/words as possible. Maybe this is just the way "I work" but then again maybe John knows that as well. I won't even try to imagine the how or why of that statement. So he comes to me because *I was open that day? * I had a connection to him in the past? * He knows I'll try to get the message across to you, thru Julie? It could be any reason. You can ask Julie or Gary. I often get "drop-ins" and friends from past readings . . . . it's almost like the connection still "holds" . . . . So, why he came to me and not you? Maybe he did come to you but you weren't really "sure" at the time or you weren't paying attention, so then why not have it come thru a way that will give an "impact" to his short statements?

Now to the reading . . . .

Bill asked:

1. When you said "Also did John's father recently bump his head on the right front side?" can you describe the image, if there was an image, that led you to make this comment? If it was an image, was it one you have of my son John or some other person? I am curious to know if it might fit my son-in-law who actually hit his head the night before or if it was perhaps John showing you something using his image."

OK, once I figured out it was John (Casper) he gave me some info . . . then I suddenly felt like I bumped my head and I felt it on the right front side. I knew John was trying to "impress an energy" to me that someone bumped their head. I thought it was for you Bill because I asked Julie to send YOU the email. It felt male. Since I had You and the bump and a male feeling, I said you. (This is where I learn from a reading and I'm sure Gary is shaking his head about now. He would ask why didn't you just say a male connected to Bill?) Well, . . . hm . . . . maybe because in this case I "assumed" yes, I just used that word, the logical? Male . . . Bill . . . bump on his head . . . the right side . . . Eureka! Bill bumped his head on the right side. OK, so I need to work on that area. HOWEVER, that also helps me for next time. I can say I do work toward getting it exact. I was close. It was a bump on the right side and it was related to you Bill, however it wasn't you. BUT!!!! The message did get across and it was received! So in one way it did in fact work. As for the science side . . . . I'm working on it.

Bill asked:

2. You made comments about "a birthday or congrats but I am seeing a celebration coming from him", and "I also feel like someone recently joined him that he knows because he is making me feel like someone "new" is with him".

OK, here it's a little easier but tricky as well. I already established I know John is coming thru. So when I see a young man holding a cake with candle I KNOW it's a celebration. (That I saw as a vision.) It seemed positive and I felt his smile. So it was a happy occasion. It's almost as if John was showing me props and at the same time making me feel. So I had two things coming from him. Visions and feelings. Now I couldn't tell you what he looked like exactly, but I did have a hair impression at the time. Like he had nice hair or hair I would like, kind of wild or a little longer. No clue why I would get that. I also saw a zipper . . . . so that lead me to believe he had a jacket on. BUT to me the message that he was sending was the importance of the quick reading. He had a message so by golly I better get hopping and get it!

Now the cake disappears that quick and he then shows me . . . . a "new person" standing next to him, who was shorter than him. I can't say I saw the person exact, but I "knew" they recently passed. Now, how do I know this? By the "energy" I was feeling. It wasn't a person who's been there for a while. I guess you could compare it to a "new employee." It often takes a while for them to fit in the groove? Or get adjusted to a new place? However, he wanted me to let you know he knows what's going on. It's that simple and that complicated!

Bill asked:

3. Can you describe the way in which you sensed these messages? Were they images, just feelings, a combination of the two or something else?

Does my explanations help? You could look at it as he gave each of you (Your daughter, your wife and yourself) a "little piece" of himself. For being a short impromptu reading, he said a lot in a little.

I would like to add . . . . I am honored to work with Gary and Julie on trying to bridge the science and the afterlife. Every step IS a step toward learning and creating a new possibility to communicate, or a new experiment. It makes me wonder how capable and how much more can I expand my abilities beyond what I already have? On the same note I want to bring something concrete to the sitter and when I do, I'm happy. When I'm off, I wonder why? as well as wonder how can I look at what I saw and understand it better? So, it's always a learning experience.

As far as the testing mediums website . . . . well, I went searching wondering what it was. I found it and signed on. Honestly, I don't think it's for me, but great there is a forum for science and scientific thoughts on the Afterlife. I'll let everyone dissect everyone's readings and pick, poke holes in and do whatever they wish to do to it "for the better" of next time. I think I'd rather wait until Gary or Julie says . . . . Hey Janet, we're working on a new experiment, do you want to try? Or if they just want me to try something new or pick my thoughts about my ability I'm up for it. (you too Bill!)

But I rather not read a lot of negative input, no offense . . . . I know what I see is REAL, and how can

I prove it to someone else? I can't MAKE someone see something, I can only share what I receive and hope that's enough! However, on that note, when you have grown up seeing, hearing and feeling things that supposedly "don't exist" you realize what a close-minded and narrow world this is, as well as how much people are missing. At age five, I wasn't told WHAT to believe when I saw my dad standing in the doorway and I could see THROUGH him. I KNEW what I saw. It was that simple and that scary. It couldn't be explained to me at the time because I already SAW it and no one else did. However by the time I realized what I did see, it was too late . . . . my new "beliefs" had already been altered and established to what "is possible and what's going on out there" not what I was told to believe. That's the difference. So although I'm happy scientists are always trying new ways and I'm happy there is a forum for thoughts on science and the afterlife, I don't think it's right for me. If my husband wants to read it, fine, but it's not for me. I hope one day it can be established through Gary's work and the work of others that it is possible, what "we" see, hear and feel does exist. You can label it psychic, telepathy, ESP, Mediums, ... Whatever, but when you see them, they ARE there! Unless someone else sees them how can we explain that to a non believer? That's why people like you, Gary and Julie are here . . . . to try to help create a process that others can understand?

Besides when Susy came to me, it blew me away. She PUT me in this path, literally. Can you imagine me emailing a man I have never met and telling him about a woman who came to me and said she was Susy, and that she died? Talk about going out on a limb! Thank Goodness for me, Susy knew what she was doing by sending me to Gary!

Bill said:

> Thanks again for caring enough to send your messages.

> No problem! Hope I didn't ramble on too much? I have a habit of doing that . . . .
>
> Janet

From my perspective, I don't think Janet could have done a better job of describing mediumship—not just the sensations a medium experiences, but also how different their world is than that of most scientists. I sometimes think I would be better off if I could just accept things at face value the way she and others do, but I wasn't given that option.

As I read through this email again recently, it finally occurred to me what the comment about the hair and the jacket with zipper were probably about. Becky's husband Aaron was the guy who bumped his head, and now that I think about it, those two comments could apply to him. Aaron rides a motorcycle and wears a leather jacket with a zipper when he does. The hair comment could also apply to him, but that is more subjective. I think John was trying to give Janet a better image of Aaron than just the bump over his right eye. This is a good example of something I pointed out previously; the sitter does not always get the significance of a comment when they first evaluate a reading, but it may occur to them later.

In her book *Spirits—They Are Present* Janet uses this reading as an example of how loved ones are often nearby and watching over us. I feel very fortunate to have been able to receive readings from Janet. It seems to me that she has a special connection to my son John. I would really recommend that anyone genuinely interested in understanding what it is like to be a medium, as well as to get a better understanding of what can be learned through mediumship, read her book.

I recently called Pat McAnaney, the psychic that Diane and I met with soon after John's death. It has been over twenty years since Diane or I talked to him, so I didn't expect him to remember us. He not only remembered both of us—he says he still remembers the image of John standing there smacking himself in the forehead saying "stupid, stupid, stupid", and that his (Pat's) dilemma at that time was trying to come up with a gracious way to relate this to Diane and me. He also said he remembers John because of the sense of humor he displayed. I think this is another example of how real these situations are to the psychic/medium, and how it may not always be easy to present what they are receiving to someone like grieving parents.

Pat explained that mediumship—communicating with the deceased—is not his strong suit and he does not usually give readings the way other mediums do. He said that the day Diane and I met with him, when Diane mentioned John's name "John was just *there*—and it all seemed apparent."

Pat's ability is in reading people's auras and using their auras to help predict their behavior and the consequences of that behavior. I told him that I really didn't understand auras and asked if he could briefly explain the difference between an old and a young soul to me. Pat graciously obliged.

"Old souls" are usually people who have a broad range of interests (scholar/athlete/creative/scientific). He says that it is always fun watching an old soul trying to pick a college major—they want to do everything. They can also be somewhat overly serious or overly responsible as children. The classic old soul is also able to step across cultural, racial, age and gender lines a little more readily than the younger souls.

Really "young souls" can be driven, ambitious, fanatic and *love* technology. The world's best gymnast, violin player or mathematician is usually a younger soul. They tend to be able to focus on one thing to the exclusion of all else. They can also be very religious and very naïve when they step outside their area of expertise.

CHAPTER 10

Fortune Tellers, Frauds or Gifted?

Trust not too much to appearances.

Virgil[5]

Once I began to realize that there is more to "life" than what literally meets the eye, my perceptions of life began to change. The first thing that became obvious is that our physical bodies, using our five senses, have very limited capabilities to sense what is actually around us. Our senses work great for the physical part of our existence, but the people we call mediums, or psychics, apparently really do have a "sixth" sense. To me this capability is analogous to some of the physical phenomena we use, such as radar, sonar and infra-red sensing to detect things we can not actually detect with our five senses.

One of the attitudes I encountered when first talking about the subject of mediumship with people is that there is a perception by many that most, if not all, psychics are frauds. While there are frauds masquerading as genuine mediums, it certainly doesn't mean that all mediums are frauds, any more

---

[5] Virgil (October 15, 70 BC-September 21, 19 BC) was an ancient Roman poet of the Augustan period.

than any other group can be characterized by the bad apples in that group. In addition, there are mediums who may have genuine paranormal abilities but who take advantage of them for personal gain, thereby also giving a bad name to the group.

As explained in Ch. 9, it can be difficult for a medium to interpret and put into words the sensations they are receiving. In any area of human endeavor, for example the improvement of artistic or athletic skill, no matter how much talent a person has they must still put in a good deal of time and effort to improve their skill level. There is a learning curve involved, and it usually requires some sort of coaching. The same is true in the development of mediumship skills. As Bob Ginsberg of the Forever Family Foundation pointed out, many people who enter the field of mediumship are well-intentioned, but are underdeveloped and unable to do what they claim. The high profile mediums, combined with the proliferation in the media of ghost and medium shows, have spawned a new industry of medium "wannabes."

This is one of the reasons that organizations doing afterlife research, such as Julie Beischel and Mark Boccuzzi at the Windbridge Institute, or those involved in helping the bereaved, such as Bob and Phran Ginsberg at The Forever Family Foundation, have medium certification programs.

The following is taken from the Windbridge Institute website:

> "The Windbridge Institute is an independent research organization consisting of a community of scientists of varied backgrounds, specialties, and interests investigating the capabilities of our bodies, minds, and spirits and attempting to determine how the resulting information can best serve all living things. Members of an extensive International Windbridge Scientific Advisory Board (SAB) review

research questions and protocols and provide feedback and suggestions based on their expertise."

Many of the research projects done by the Windbridge Institute involve the use of mediums, so it is important that the mediums used in their studies meet minimum standards of performance. Because of this requirement, the Institute has developed it's own medium training and screening process. For those who may be interested, this process is described in a paper by Dr. Beischel. The Windbridge Institute website address is given at the back of this book. (In a recent communication with Dr. Beischel, I learned that they are not screening additional mediums at this time).

Dr. Beischel's new book, *Among Mediums: A Scientist's Quest for Answers*, was published on Jan. 15, 2013. It is available in Kindle version only at this time. In this easy-to-read and entertaining book, Dr. Beischel answers many of the questions people have about the validity of mediumship. She explains the protocols used to evaluate potential mediums and discusses her research methodologies for determining the accuracy and specificity of information reported by mediums. This book is an excellent example of how science is being used in the investigation of paranormal phenomena.

The Forever Family Foundation (FFF) also has a medium certification program. The main purpose of the FFF, as stated on their website is "To further the understanding of Afterlife Science through research and education while providing support and healing for people in grief." It further states that one of the objectives is "to provide a forum where individuals and families who have suffered the loss of a loved one can turn for support, information, and hope through state-of-the-art information and services provided by ongoing research into the survival of consciousness and afterlife science."

In order to provide a reliable resource to which the bereaved can turn the foundation provides the names of mediums who have passed their certification process. The FFF website address is also given at the back of this book.

I am going to list several points that I hope will give the reader a clearer picture of what mediums are like, and what is reasonable to expect from them.

- I have received readings from several outstanding mediums but I have never seen one who wears strange clothes while looking into a crystal ball.

- Mediums are people too and no one is perfect. Don't expect them to have all the answers or get it right every time.

- Each medium is unique. However, they must work within their own frame of reference. For example, two mediums may get the same vision but interpret it differently.

- Each medium's area of expertise is a little different than the next—just as people in other walks of life have various types of skills and levels of ability. One medium may see visions and another may hear spirits speak to them. One is not better than the other.

- Some mediums get names—some don't. Some get dates—some don't. Some lean more toward personality—how the spirit lived while they were incarnate.

- The medium has no control over which discarnate may come through. That seems to be up to the discarnates.

- Many mediums may connect differently with different spirits. When someone dies they don't lose their personality. Think of someone you get along with. Now think of someone you don't get along with. Not every medium meshes with each spirit.

- Some spirits come through stronger than others. A shy spirit may not come through as strong.

- When someone dies they are not "all knowing." They don't suddenly have all the answers.

Here are a few suggestions for those who may want to receive a reading:

- Try to be open minded—but objective. I'm going to give a couple of actual examples of people being close-minded. 1) The medium was being shown what appeared to her to be a wake board or a slalom water ski. The sitter told her she was wrong because her loved one died in a snow boarding accident. 2) The medium told the sitter that she was being shown what appeared to her to be something related to a large food business. The sitter told her she was wrong because her family owned a chicken farm—not a large grocery store.

- When receiving a message—at no time will a spirit tell you to do something that you are in conflict with.

- Never provide the medium with information prior to a reading. If the medium makes a statement, a "yes" or "no" answer is often all you need to give. Clarifying the facts after the medium makes a statement is fine.

- The information provided by the medium should be specific.

- Be careful if a medium wants to have a deep "conversation" with you—they may be fishing.

- When looking for a medium, ask someone you can trust or check out credible organizations.

There are an increasing number of organizations dedicated to research in the area of paranormal phenomena—organizations such as the Institute of Noetic Sciences (IONS). This institute was co-founded in 1973 by former astronaut Edgar Mitchell and investor Paul N. Temple to encourage and conduct research on noetic theory and human potentials. Institute programs include "extended human capacities, integral health and healing and emerging worldviews." (taken from Wikipedia). A Google search will reveal many organizations investigating a wide variety of paranormal phenomena.

I find it encouraging that paranormal phenomena in general, and mediumship in particular, are gradually being taken from a position of "occultism" to a place where people with legitimate psychic abilities are accepted within the scientific community.

CHAPTER 11

What conclusions can we draw from this?

True wisdom comes to each of us when we realize how little we understand about life, ourselves, and the world around us.

Socrates

At first, because I had no concept of what the "afterlife" (or hereafter, or Heaven, or whatever name you prefer) might be like, assuming something of this nature exists, I was mystified by what the mediums were able to do. My entire world at that time was defined by what I had been taught, what I read or was told by various forms of media and what I could perceive with my five senses. It consisted of only the physical realm.

All of a sudden I was shown something very different, and I began to try to understand it. After going through the first nine readings the day I received them, I asked myself if it was possible that Dr. Schwartz had somehow been able to find out all this information about my family. I concluded that he could have found out some of it, such as my dad having lived in Hawaii, my mother dying years before my dad, etc.

But how would he have been able to find out about my dad having a serious problem with his throat, or that he was a world traveler, was religious, grew up in poverty and so forth? How could he find out that Diane's father had been a heavy smoker and was bald with spots on his head?

First of all, I had just met him the day before I received the readings. We had very little contact prior to that day, and I could come up with no reason he would even want to go to the trouble of doing such an extensive search. We had not even discussed any kind of a relationship yet, and I had only expressed an interest in his work.

Secondly, when I met him at his hotel that first day, I told him about losing my son John, but said nothing about my dad or Diane's parents. I had been hoping to hear from John, so I think it is logical to assume that if he were going to do such an extensive search, he would have done research about John but not my dad.

I finally concluded that he would need a better search organization than the FBI and CIA combined. Not even Google was in wide use at that time. In addition, he did not fit the profile of a person that I would think of as a scam artist. He was a university professor doing research and was publishing his work for everyone to see. It didn't take long to rule out some sort of scam.

Something that happened while we were doing one of the first Asking Questions experiments helped me to accept and also to understand a little better what mediums do. Laurie Campbell, the medium, was seated in one room and the sitter was seated behind a divider in an adjacent area out of Laurie's sight. Laurie had been given a list of questions prepared by the sitter. I was sitting in a separate room, also out of sight, but within hearing distance of the other two. The sitter's mom had passed recently and she was hoping to be in contact with her. One of the sitter's questions was "what favorite meal did mom used to prepare?" Before the medium replied, an image of some green vegetables appeared in my head. A few seconds later, Laurie gave a similar answer, and the sitter said that was correct. I was really surprised. It is often said that everyone has some psychic ability, but that was the first time I had experienced anything that suggested I might have a small amount.

Recently, I had another experience that I can now accept as possibly being caused by some outside "influence", as opposed to it being a "coincidence." I have an outdoor aviary divided into three sections that is large enough to walk into and that is located down near the creek on my property. I started it soon after losing John. At this time I have about twenty Parakeets and a few Canaries. I also have, in addition to deer, raccoons, skunks, rabbits, wild turkeys, snakes, gophers, possums, mountain lions, coyotes, etc. that live in the area, quite a few rats who nest in the ivy growing all over my property. About fifteen years ago the rats chewed a hole through the wood framework supporting the galvanized wire screening, went in and killed all fifteen of the finches in one section, and a couple of weeks later chewed a hole into another section and took twelve out of the thirteen canaries I had at that time. I spent days cementing the bottom, re-enforcing the screening, etc, and did not have another problem for about fifteen years.

Then several months ago I discovered that a rat had gotten in by gnawing through a corner of the concrete flooring, was nesting inside (hidden under a wood platform in one section) and was picking off my birds one or two at a time every night. I got rid of the rat, put wire mesh in the hole and re-cemented it. Several nights later I was sound asleep but suddenly woke up about 3am thinking and worrying about my birds. I had this sense that a rat was getting into the cage. At first I told myself to go back to sleep, but I couldn't get rid of the concern, so I got out of bed, put on my pants and shoes, grabbed my flashlight and went down to the cage. Just as I opened the cage door, a large rat that was in the cage by the door ran away from the door in a panic and began dashing back and forth. These guys are about a foot long (not including tail) when full grown, and can definitely see better in the dark than I can. It would run to near where I was standing, but when I shined the light on it, it would retreat to the back of the cage, run up the side, then come running back. After repeating this quickly about four or five times, it ran back to where I was and just disappeared! I searched around the cage, but no rat!

I checked my birds and they were all OK. The rat didn't have time to grab one of them, so I apparently got there just in time. Then I got down on my hands and knees in the area where it had disappeared and began checking the wire screening. After a few minutes, I found the hole where it had gotten in and then escaped through. I figured it must have been a rusted or damaged section of screen. I spent the next half hour or so patching the hole in the dark and then went back to bed, happy that I had averted another night of horror for my birds. In order to be certain that it was not a dream, I went down to the cage the next morning and checked my patched up hole to be sure it was secure. It is still there.

It's possible that because of past rat problems there was some subconscious concern in my brain that woke me up. If this was the case, why did these concerns surface at the exact time they did? Did I just "happen" to wake up and go down to the cage *just in time* to save my birds, or could I have received a "telepathic" message, warning me about my birds and the rat, that caused me to get up and go do something about it? If I did receive a "message", where did it come from? On the other hand, if there was some outside influence involved, why did it not show up on previous occasions? One answer may be that I was just not open to it.

CHAPTER 12

Memory

Memory is deceptive because it is colored by today's events.

Albert Einstein

By the time we had finished the "Asking Questions" experiments, I was not only becoming convinced that we survive physical death but I was beginning to get an idea of what life following physical death may be like. Questions began entering my mind: If we survive the death of our physical bodies, what is it that actually survives? Is it our mind? If it is our mind that survives, does it contain our memories? What evidence is there to support this concept?

Some of the books on paranormal phenomena that I read in the early days following John's death mentioned studies conducted by Dr. Ian Stevenson at the University of Virginia who did extensive research on the subject of reincarnation involving children.

The following is taken from Wikipedia:

"Ian Pretyman Stevenson, MD, (October 31, 1918-February 8, 2007) was a Canadian biochemist and professor of psychiatry. Until his retirement in 2002, he was head of the Division of Perceptual Studies at the University of Virginia School of Medicine, which investigates the paranormal.

Stevenson considered that the concept of reincarnation might supplement those of heredity and environment in helping modern medicine to understand aspects of human behavior and development. He traveled extensively over a period of 40 years to investigate 3,000 childhood cases that suggested to him the possibility of past lives. Stevenson saw reincarnation as the survival of the personality after death, although he never suggested a physical process by which a personality might survive death."

Dr. Stevenson cites two primary factors as possible evidence of reincarnation. These factors were noted during his investigation of children claiming to remember events from one of their past lives: 1) birthmarks and birth defects corresponding to wounds on deceased persons whose life the child claimed to remember and 2) the ability of some of these children to speak a language they did not learn by normal means. This is known as Xenoglossy.

The significant point here is that, if indeed reincarnation is an actual phenomenon, and people can remember past lives, it implies that memory survives the death of the physical body. This is a common belief in some Eastern religions. My purpose here is to introduce the concept. For those who are interested, there is a great deal of information available on the internet about Dr. Stevenson's work. I have listed three websites at the back of this book.

The other concept that supports the hypothesis of memory survival following death is that of past life regression. I described my experience with this phenomenon in Chapter 5. Recently, Diane asked me again if I wanted to read the book on past life regression that she had mentioned years ago, and I again told her no, but a few days later it occurred to me that if this phenomenon is real, it also has a very important place in the afterlife research area. It suddenly dawned on me that if there is anything to these two phenomena which have major implications related to our memories, then I needed to learn more about it. I read the book and found it very interesting. For those who may be interested, the name of the book is *Many Lives, Many Masters* by Brian L. Weiss, M.D.

The key point here is that the concepts of reincarnation and past life regression conflict with the theory put forth by many mainstream scientists that memories are stored only in the brain.

There are three general categories of memories that I am aware of:

1. Our personal memories

2. Physical memories such as those stored and passed on from one generation to the next by our DNA.

3. The Akashic Records including, but not limited to, the collective unconscious as defined by Carl Jung.

The memories I am discussing are our personal memories. The following definition of memory is taken from Wikipedia:

> "In psychology, memory is the processes by which information is encoded, stored, and retrieved. Encoding allows information that is from the outside world to reach our senses in the forms of chemical and physical stimuli. In this first stage we must change the information so that we may put the memory into

the encoding process. Storage is the second memory stage or process. This entails that we maintain information over periods of time. Finally the third process is retrieval. This is the retrieval of information that we have stored. We must locate it and return it to our consciousness. Some retrieval attempts may be effortless due to the type of information."

If our personal memories *are* stored in our brain while we are alive, which has never been proved or disproved, then the options are:

1) Because the brain is a physical object and dies with the rest of the body, the memories simply die with the person. This implies there is no afterlife.

2) If these memories still exist and can be recalled in a subsequent life following a person's death, it implies that there is a *before and after* life, and the second option for these memories is that they were transferred out of the brain at death.

However, there is a third possibility, which is that these memories were never stored in the person's brain, but instead were stored outside of the brain. Science based theories supporting this option are discussed in the next chapter.

CHAPTER 13

A Scientific Look at Paranormal Phenomena

Science is a way of thinking much more than it is a body of knowledge.

Carl Sagan

A couple of months after losing John, I was telling a friend that I was determined to find out if John survived the death of his body. He recommended that I read a book that had recently been published entitled *The Presence of the Past* by Rupert Sheldrake, who has a PhD in biochemistry from Cambridge University. The following is taken from the Amazon website:

"Rupert Sheldrake's theory of morphic resonance challenges the fundamental assumptions of modern science. An accomplished biologist, Sheldrake proposes that all natural systems, from crystals to human society, inherit a collective memory that influences their form and behavior. Rather than being ruled by fixed laws, nature is essentially habitual. *The Presence of the Past* lays out the evidence for Sheldrake's controversial theory, exploring its implications in the fields of biology, physics, psychology, and sociology.

At the same time, Sheldrake delivers a stinging critique of conventional scientific thinking. In place of the mechanistic, neo-Darwinian worldview he offers a new understanding of life, matter, and mind."

This was the first book I had ever read that questions some of our fundamental scientific beliefs. It is one of the reasons I began thinking differently about our world, and to question many of the concepts that I had been taught and had accepted as fact.

I am going to list a few of the key points made by Sheldrake

1. In the introduction, he points out that, until as recently as the 1960s, scientists believed that the universe was essentially static. It was believed that it's properties, fields, and even the laws of nature were unchanging. We now know that the universe is a very dynamic, constantly changing environment. (Sheldrake, 1989, pg. XIX).

He also states in his introduction: "Any new way of thinking has to come into being in the context of existing habits of thought. The realm of science is no exception. At any given time, the generally accepted models of reality, often called paradigms, embody assumptions that are more or less taken for granted and which easily become habitual." (Pg. XX).

2. Because his book is introducing a new concept, something which he calls morphic resonance that theoretically involves morphic fields, he discusses various types of fields, such as gravitational and electro-magnetic fields, and states:

"Fields are non-material regions of influence. The earth's gravitational field, for example, is all around us. We can not see it—it is not a material object, but it is nevertheless real."—"There are also electro-magnetic fields, which are quite different in nature from gravitation. They have many aspects. They are integral to the organization of all material systems, from atoms to galaxies. They underlie the functioning of our brains and bodies."

(Sheldrake, 1989, pg. 97).

3. He uses the analogy of a TV set to point out that the images seen on the screen are the result of the transmission of coded signals, via electro-magnetic waves, which are then decoded by the components in the TV set. In other words, as most people are aware, the images are not actually stored in the components inside the set. (Sheldrake, 1989, pg 136).

4. In Ch 12, *Minds, Brains and Memories*, he makes the point that science has never been able to show that our memories are stored in our brain, and he raises the question of whether or not they could be stored in some sort of field, analogous to the way TV pictures are stored in the electro-magnetic field waves that are transmitted from the TV station and picked up by the TV's antenna.

The four main points that I am emphasizing, therefore, are:

1) science is constantly forced to restate it's principles as we gain new understanding of the physical world.

2) there are things that we can not sense, but that we know exist and that have a profound effect on the physical world, including our bodies.

3) we know that information is stored in and transmitted by invisible energy fields.

4) science has never been able to show that memories are stored in our brain, or for that matter, that our thoughts originate there.

There seems to be a common belief, both among the general public as well as among many highly educated people such as doctors and scientists, that all of our thoughts and ideas originate in our brains. They make no distinction between the mind and the brain, and often tend to use the two words interchangeably.

A few years after reading Sheldrake's book, I was on a flight to Argentina reading a book related to paranormal phenomena and I ended up in a discussion with an engineer sitting next to me who had asked me about my book. In the discussion, I told him that I thought it was possible that our thoughts do not necessarily originate in our brain, or that at least they could be influenced by some outside, unseen source. I used the analogy of the TV set to make my point. He immediately told me how absurd he thought that was, so I jokingly asked him if he really believed all those little people in his TV set actually lived there. We chatted for awhile and it turned out we were staying in hotels near each other, so we agreed to meet for dinner. As we met at the restaurant, he suddenly sort of blurted out how my comment about the TV set made him think, and he admitted that my idea was at least possible. That was the effect I was trying to cause, and am with this book—trying to create an awareness of novel ideas that will help people to think different, even original, thoughts. I have come to believe that often fear is the result of a lack of understanding.

That incident reminds me of a conversation I had with a couple of friends recently. They were commenting about how dumb chickens were and one of them stated that the reason was because they had small brains. I asked him if that therefore meant that insects such as bees and ants, with their organized societies, were really stupid since their brains are really tiny. It would also imply that animals such as elephants and whales are much more intelligent than humans. I didn't get a reply.

I have heard that comment several times—that animals like turkeys and chickens are stupid because they have small brains. We have all heard the expression "bird brain". I am not questioning that these animals may not be intelligent, at least not according to our definition of intelligence, but rather if it is determined by brain *size*. We know that one of the primary purposes of our brains is to control the physical functions of our bodies, and from what I can tell most people believe that it also serves as the source of thinking, rationalizing and storing memories. I began to wonder if it could also be to interface with some outside "information source."

Under the title of *"The Electrical Evocation of Memories"*, Sheldrake mentions studies done by Wilder Penfield, a neurophysiologist, conducted during the course of operations on conscious patients with various neurological disorders.

"He and his colleagues tested the effect of mild electrical stimulation of various regions of the brain. As the electrode touched parts of the motor cortex, the appropriate limb movements would occur. Stimulation of the primary auditory or visual cortex evoked auditory or visual hallucinations: flashes of light, buzzing noises, and so on. Stimulation of the secondary visual cortex gave rise to complex, recognizable hallucinations of flowers, animals, people and so forth. And in epileptics, when some regions of the temporal cortex were touched, some patients recalled apparently specific memory sequences, for example an evening at a concert or a telephone conversation. The patients often alluded to the dreamlike quality of these experiences."

Sheldrake goes on to point out that Penfield initially assumed that these memories were stored in the stimulated tissue; or it could mean that stimulation of that region activated other parts of the brain that were involved in remembering the episode. Significantly, after further reflection, Penfield abandoned his original interpretation.

> "In 1951 I had proposed that certain parts of the temporal cortex should be called "memory cortex," and suggested that the neuronal record was located there in the cortex near points at which the stimulating electrode may call forth an experiential response. This was a mistake.—The record is not in the cortex."

"Penfield gave up the idea of localized memory traces within the cortex in favour of the theory that they were distributed in various other parts of the brain instead, or as well." (Sheldrake, 1989, pgs. 219-220).

The question is: Were the images seen by Penfield's patients stored somewhere in the brain, or did the electrical stimulation of brain tissue cause activation of an antenna effect between the brain and some field outside of the brain that permitted these images to be seen by the patients?

In his book *Matter to Mind to Consciousness: Anatomy of the E.L.F.*, T. Lee Baumann, M.D. points out that

> "Millions of central nerves appear to end at the surface level of the cerebral and cerebellar cortices. These nerves are not currently recognized to complete the electrical circuits that researchers have always expected. At this level of the cortex, all nerves were thought to synapse sequentially in a never-ending cycle."

He goes on to state that "Somewhere, in one of these cerebral processes, researchers believed our conscious awareness existed. Of course, with today's limited technology, this has not proved possible." He continues:

"If we allow ourselves some open-mindedness, then there are two possible explanations for why neurons at the cortical surface appear to end blindly:

1. Perhaps it is possible that every neuron does connect to another neuron—it's just that our present technology is limited, and we cannot trace and identify every neuronal connection.

2. The remaining explanation is that many of these cortical neurons do end blindly. Their purpose, then, is to function as part of a vast EM (electro-magnetic) antenna network involving consciousness, yet to be more fully described.

If the second explanation is correct, then how do these cells function?

We are aware that peripheral nerve terminals in the skin are blindly ending organelles. They serve to detect, initiate and relay peripheral sensations. I suggest that an analogous—but reverse—situation exists in the outer layers of the cerebral (and likely cerebellar) cortex."

He further states that:

"My research leads me to believe that a vast number of our neurons allow us the facility to transmit and receive EM signals in the extremely low frequency (ELF) range. If I am correct, one of the prime candidates allowing this transmission are those neurons in the superficial cortical layers of the brain that appear to have terminal dendrite endings without further connections."

(T. Lee Baumann, 2010, pgs. 97-100).

In other words, could it be that the brain serves as part of an "antenna" to receive signals, which may then activate controlling functions within the brain, analogous to the way nerve sensors in the skin detect physical sensations created by external stimuli such as heat or pressure? Could it also possibly be used to send signals, such as our thoughts, to this external source/receiver?

In his book *"Where God Lives"*, Melvin Morse M.D. suggests that the actual part of the brain that performs this function is the right temporal lobe.

Another interesting article that addresses the question of whether or not consciousness is created by the brain was written by Dr. Gary Schwartz as part of one chapter in a book entitled *The Oxford University Press Handbook of Psychology and Spirituality*, published in 2012 and edited by Lisa Miller, PhD, from Columbia University.

Under the heading "Does Consciousness Require a Brain?" Dr. Schwartz points out that there are three types of evidence that together *seem* to point to the conclusion that consciousness is created by the brain. I am going to quote Dr. Schwartz.

"The three kinds of evidence are as follows:

1. *Evidence from recordings*—Neuroscientists record brain waves (via electroencephalograms [EEGs]) using sensitive electronic devices. For example, it is well known that occipital alpha waves decrease when people see visual objects or imagine them.

2. *Evidence from Stimulation*—Various areas of the brain can be stimulated using electrodes placed inside the head or magnetic coils placed outside the head. For example, stimulation of the occipital cortex is typically associated with people experiencing visual sensations and images.

3. *Evidence from ablation*—Various areas of the brain can be removed with surgical techniques (or areas can be damaged through injury or disease). For example, when areas of the occipital cortex are damaged, people and lower animals lose aspects of vision.

The generally accepted—and seemingly common sense—neuroscience interpretation of this set of findings is that visual experience is created by the brain.

However, the critical question is whether this *creation of consciousness* explanation is the *only* possible interpretation of this set of findings. The answer is actually no. The three kinds of evidence are also consistent with the brain as being a receiver of external consciousness information (Schwartz, 2002, 2005, 2011).

The reasoning is straightforward and is illustrated in electronics and electrical engineering. Though it is rare to discuss an electronics example in the context of a psychology monograph (especially one focused on religion and spirituality), it turns out to be prudent and productive to do so here.

Consider the television (be it analog or digital). It is well known—and generally accepted—that televisions work as *receivers* for processing information carried by *external* magnetic fields oscillating in specific frequency bands.

Television receivers do *not create* the visual information (i.e., they are *not the source* of the information)—they *detect* the information, *amplify* it, *process* it and *display* it.

Apparently it is not generally appreciated that electrical engineers conduct the same three kinds of experiments as neuroscientists do. The parallel between the brain and the television is essentially perfect.

1. *Evidence from recordings*—Electrical engineers can monitor signals inside the television set using sensitive electronic devices. For example, electrodes can be placed on particular components in circuits that correlate with the visual images seen on the screen.

2. *Evidence from stimulation*—Electrical engineers can stimulate various components of the television using electrodes placed inside the television set or magnetic coils placed outside the set. For example, particular circuits can be stimulated with specific patterns of information, and replicable patterns can be observed on the TV screen.

3. *Evidence from ablation*—Electrical engineers can remove various components from the television (or areas can be damaged or wear out). For example, key components can be removed and the visual images on the screen will disappear.

However, do these three kinds of evidence imply that the *source* or *origin* of the TV signals is *inside* the television—that is, that the television *created* the signals? The answer is obviously no."

Dr. Schwartz then points out that "the three kinds of evidence, by themselves, do not speak to (and do not enable us to determine) whether the signals—the information fields—are:

(1) coming from inside the system (the materialistic interpretation applied to brains), or

(2) coming from outside the system (the interpretation routinely applied to television).

It follows that *additional kinds of experiments* are required to distinguish between the "self-creation" versus "receiver" hypothesis.

Experiments on the survival of consciousness (SOC) hypothesis with skilled research mediums provide an important fourth kind of evidence that can neither be predicted nor explained by the self-creation (i.e. materialism) hypothesis, but it can be predicted and explained by the receiver hypothesis." (Schwartz, 2002, 2005, 2011).

As I began to develop an idea of what the afterlife may be like, and became aware that our deceased loved ones seem to be around us, it occurred to me that they may be influencing us to a much greater extent than I would have thought possible. Questions began to pop into my head, such as: To what extent can they influence us? Can they put thoughts in our minds? When I say something like "It occurred to me", or when an inventor thinks of a new idea, where do those thoughts come from? Is it really the product of some random firing of neurons in our brain, or is there a more logical explanation. Do all of our thoughts originate strictly in our brain, or is there some other influence involved?

CHAPTER 14

Materialism versus Dualism

A cold, atheistic materialism is the tendency of the so-called material philosophy of the present day.

Adam Sedgwick[6]

I think my introduction to the idea of dualism also came from Rupert Sheldrake's book, *The Presence of the Past*. He made me aware of the concept of monism versus dualism. The monism I refer to, also known as "physicalism" or "materialism", holds that only the physical is real, and that the mental or spiritual can be reduced to the physical. Discussions of monism versus dualism have been going on for thousands of years, most notably in ancient Greece, and the concept of dualism underlies, for example, the customs practiced by the ancient Egyptians when burying their Pharaohs. However, with my materialistic view of the world I thought of their culture as just an ancient belief system that certainly didn't apply to today's world.

---

[6] Adam Sedgwick (March 22, 1785-January 27, 1873) was one of the founders of modern geology. Though he had guided the young Charles Darwin in his early study of geology, Sedgwick was an outspoken opponent of Darwin's theory of evolution by means of natural selection.

I am going to quote from Sheldrake;

"Materialists believe that the mind is in the brain. One form of materialism treats conscious mental activity as an epiphenomenon of the activity of the brain, rather like a shadow; the conscious mind is an effect of the physical activity of the brain, but it is not a cause; it has no function at all, and everything would go on just the same without it. Another form of materialism asserts that conscious mental activity and brain processes are simply different aspects of the same reality."

(Sheldrake, 1989, pgs. 210-211).

A few months after reading Sheldrake's book I enrolled in an adult education philosophy class at Stanford. The class mainly studied the philosophies and ideas of two people—Descartes and Jung. I learned that Descartes was not only the founder of the Cartesian coordinate system used in mathematics, but that he was also a great philosopher. I now understand why his statement "I think, therefore I exist" is considered so profound. It implies dualism. The dualism I refer to is termed Cartesian Dualism, i.e., mind-body or mind-matter dualism and it states that the mind is *separate* from the body.

The way I think of these two separate parts that make up, for example, one person (or animal, insect, etc.) is that the physical part is merely a "housing" adapted for use while on Earth—the only place, as far as we know, that it can function unless we are able to bring a "life support system" with us. As we age, this physical part gradually deteriorates and eventually fails to function properly. We call this death.

The second part, our mind, is more difficult to define. The term generally used in psychology is "consciousness", a term that refers to the relationship between the mind and the world with which it interacts. It defines our ability to be aware of ourselves and what is around us, and has been the subject of debate at least since the time of Descartes. The problem I have with the definition above is that, to me, it has too much of a "mechanistic" tone to it. It sounds as if we are some sort of "object" that senses what is around us and responds accordingly, like a robot.

I prefer to think of the mind as essentially our emotions—love, fear, hope, hate, anxiety, happiness, anger, excitement, compassion, curiosity, disgust, and so forth—in other words, our *feelings*. These are not physical components of the body. You can not touch them or put them in a box, but you can feel them yourself and observe them in others. In describing a person, we often describe them in both a physical and an emotional sense—for example—"she's very pretty but seems to be unhappy."

It should be noted that millions of people are dualists. Most practicing Christians and Muslims, for example. believe in a hereafter, and it seems that modern science, in far too many instances, treats this as absurd.

In his book *Visions, Trips and Crowded Rooms*, David Kessler, a renowned expert on death and grief, relates three experiences that the dying often undergo:

1. "Visions." As the dying lose sight of this world, some people appear to be looking into the world to come." Diane witnessed this while sitting with a dying friend, who began talking to someone who was not visible to Diane. The now-famous last words of Steve Jobs, as related by his sister, were "Oh wow. Oh wow. Oh wow." According to an article in the Washington Post, he was looking past his family members (over their shoulders) when he uttered those words. Could he have had a vision of something fascinating? I would think that if it had been a horrible vision, he would have said something like "Oh no!" Of course, there

are other possible explanations, but a vision certainly seems like a good possibility.

2. "Trips". The dying often think of their impending death as a transition or journey." This particular item reminds me of the conversation my wife had with our son John the night before he was killed. This is taken from her notes:

May 1, 1991.

"That night John and I had a long and amazing phone conversation. He told me that he was having macaroni and cheese for dinner . . . . not so amazing, but something I'll never forget. He told me how well he was doing in school (on the honor roll!) and talked about plans for the next semester. He then went on to say, "I have a very strong feeling that I'm going to take a trip . . .", but sounding somewhat confused, said that was strange because he was in no position to be going anywhere because he was doing so well in school and planned to finish at Cuesta College so that he could transfer to a four year college as a junior. John kept going 'back and forth' repeating this "feeling" that he was going on a trip. Finally he said, "Maybe I'll call one of my friends and ask them if they have something planned." I feel that John had a very strong, but probably subconscious premonition that he was leaving, but it confused him as it didn't make "sense."

3. "Crowded Rooms." The dying often talk about seeing a room full of (discarnate) people. We never die alone." (This reminds me of the first reading by Janet Mayer where she said that there were several people waiting to "come through").

In his book, Kessler gives examples of numerous occasions where people have these experiences shortly before they die. The people relating these incidents (those who witnessed the dying having these experiences) are various professionals, including doctors, nurses, social workers, counselors, psychologists and chaplains primarily working in hospices, hospitals, nursing homes, palliative (end-of-life) care facilities and so forth.

Interestingly, while there are many doctors who have witnessed and acknowledged people having these visions as they near the end of (physical) life, far too many still claim that these visions or other death bed experiences are hallucinations due to drugs being given. If they are informed that the patient is not on a hallucinatory drug, they then often attribute it to a lack of oxygen. Because of their biases, they are willing to attribute it to anything other than a vision that a large number of people experience just before dying.

In the Epilogue, Kessler cites a study done at the Camden Primary Care Trust in London regarding the believability of deathbed visions.

> "Interviews revealed that patients regularly report these phenomena as an important part of their dying process, and that DBP (deathbed phenomena) are far broader than the traditional image of an apparition at the end of the bed. Results of the interviews raise concerns about the lack of education or training to help palliative-care teams recognize the wider implications of DBP and deal with difficult questions or situations associated with them. Many DBP may go unreported because of this. Results of this pilot study also suggest that DBP aren't drug-induced, and that patients would rather talk to nurses than doctors about their dying experiences."

(Kessler, 2010, pg. 152).

As Kessler points out, you don't hear these stories from patients who are ill but not dying. With very few exceptions, these visions only occur when someone is clearly close to death. Pg. 1.

He also points out that these experiences are different than near death experiences, in which a person survives clinical death. He goes on to say "In terms of the overall ways in which society views what cannot be easily understood or proven, there may be an unintended arrogance in the judgment of the visions presented here." Pg. 4.

He makes some other interesting points:

1) Statements made by the dying are admissible in court legal proceedings, and a person can be convicted and sent to prison based on this testimony. Pg 16.

2) Death bed visions have been a part of the arts dating back to Shakespeare, and continue to be so into modern times. Ch 4.

3) He states that "Interestingly, I've never come across anyone who experienced a vision outside of his or her faith—in other words, a Jewish person doesn't see Jesus, and a Christian doesn't see Allah." (Kessler, 2010, pg. 89).

I find this last observation interesting because it implies a possible answer to one of the questions I had when involved in the Asking Questions experiments: Are there religious groups similar to those on Earth? Kessler's observation suggests that there may be religious groups in the afterlife similar to those on Earth, although another possibility is that we are able to have visions of what is familiar and comforting to us.

If I carry this one step further, I could make the case that life on the "other side" is very similar to "this side", and in fact, as has been suggested by others, the afterlife is more "real" than our physical life. People who have had an NDE say that the colors on the other side are far more vivid, that there are colors we can't perceive with our physical eyes, the music is exquisitely beautiful, and most of all the feeling of being loved is pervasive. Many don't want to return to this "side", but for various reasons are told that it is not yet their time to go. I sometimes think that our physical life may be an "experiment" of some sort, where we are in a type of laboratory being observed, and even "influenced" by those on the other side. It's as if we are trying to create a physical version of the non-physical world. Sound scary? Not at all to me. I find the thought, which is admittedly way out there at this point, rather intriguing, and in a sense exciting!

There is a form of mediumship generally known as physical mediumship, also known as "materialization", in which it is said that discarnates are able to temporarily take on a physical form for a short period of time using the mediums body, and can be seen by the people in attendance at one of these sessions. This is a controversial topic even within the groups who in general believe in mediumship. I have seen videos of these proposed phenomena, the most notable being one done by the Scole Group in Great Britain. I bring it up here because it reminds me of the deathbed visions discussed above, in which discarnates are able to make themselves visible to the dying. Could it be a similar phenomena?

The picture I'm getting of what our capabilities may be once we enter the afterlife is that we can:

1. Communicate images, sounds and physical sensations to people who have a special sensitivity.

2. Become visible to those still in the physical world, especially when they are near (physical) death.

3. Predict an occurrence and then make it happen as predicted. There are many documented cases of this. My Christmas tree ornament reading is a good example.

All of this implies an interaction between incarnates and discarnates that occurs on a continuing, daily basis.

CHAPTER 15

Skepticism

> *Dogmatism and skepticism are both, in a sense, absolute philosophies; one is certain of knowing, the other of not knowing. What philosophy should dissipate is certainty, whether of knowledge or ignorance.*
>
> Bertrand Russell

"In general, skepticism refers to any questioning attitude towards knowledge, facts or opinions/beliefs stated as facts, or doubt regarding claims that are taken for granted elsewhere. Philosophical skepticism is an overall approach that requires all information to be well supported by evidence." Wikipedia.

I have pointed out a number of times that I was very skeptical about virtually all paranormal phenomena when I was first exposed to them. When the question of survival of consciousness became so important to me, I forced myself to examine it with an open mind. I understand how those who have never lost someone they truly loved may not be concerned about survival following the death of our bodies, but that was not possible for me. One way I think of what happened to me is that I had my head jerked out of the sand.

A skeptic would probably argue that the trauma I experienced caused me to believe what I wanted to believe, and to distort information to make it fit those beliefs.

The question for me became simply: Do we survive physical death or not, and what evidence is there to support and/or refute either belief? That's what afterlife research is about, and that is why, for those who are not just "believers", we do scientific studies designed specifically to help us answer this question.

During the early days of my involvement in the study of afterlife phenomena, I became aware of people who considered themselves qualified to critique the work being done by paranormal investigators, but I did not spend much time researching the issue. From what I could gather, these individuals had a peripheral understanding of some of the claimed phenomena, but had never studied the subject in any great detail themselves. They didn't try to replicate the studies being done, or attempt rigorous studies of their own to prove their views. In fact, in presenting their arguments, some of them just ignore most of the data supporting the case for survival of consciousness and simply attempt to ridicule those who are trying to do these unusual studies. They often make the mistake of assuming that all those who are engaged in the legitimate study of afterlife phenomena use the same techniques that are used by the frauds. In this way they hope to discredit the legitimate scientific investigators. I wondered why they would even bother and the only conclusion I have come up with so far is that they are so biased against the concept of survival-of-consciousness that it is not possible for them to be objective.

Skepticism is important. It is a crucial part of any scientific endeavor. In an effort to try to understand the skeptic's position, I recently decided to review some of the literature that is available critiquing paranormal studies. I tried to find people who seemed to be doing valid, science-based critiques. The first article I found is an online critique entitled *How Not to Test Mediums: Critiquing the Afterlife Experiments* by Ray Hyman. This article was written soon after Dr. Schwartz' book was published.

Dr. Hyman is a Professor Emeritus of Psychology at the University of Oregon, and a noted critic of parapsychology. He has a doctorate in psychology from John Hopkins University. (Wikipedia).

In his introductory comments, Dr. Hyman states that Dr. Schwartz "makes revolutionary claims that he has provided competent scientific evidence for survival of consciousness and—even more extraordinary—that mediums can actually communicate with the dead. He is badly mistaken. The research he presents is flawed, and in numerous ways. Probably no other extended program in psychical research deviates so much from accepted norms of scientific methodology as this one." (Hyman, 2003, pg.1).

In referring to the research done by Dr. Schwartz in his lab at the University of Arizona where he had studied noted mediums such as John Edward and George Anderson, Dr. Hyman states that "This work has attracted considerable attention because of Schwartz's credentials and position. Even more eye-opening is Schwartz's apparent endorsement of the medium's claims that they are actually communicating with the dead." (Hyman, 2003, pg. 1).

Dr. Hyman goes on to list what he considers major flaws in Dr. Schwartz' research methodology. I agree that in some ways the experiments Dr. Schwartz first conducted did not meet the strict protocols that are normally employed in rigorous scientific studies. I have been involved in many engineering research and development projects, and we rarely got our test procedures right the first or second time. It always required ongoing improvements and refining of the procedures. If it is that difficult to do in the physical world where we can generally see, touch and control the various factors affecting our work, how difficult do you think it is when trying to develop protocols to conduct experiments in the study of something like afterlife phenomena—a science that is in it's infancy?

As I explained at the end of Chapter 7, I felt that we needed to improve the scoring by adding additional qualifying criteria. Also, Dr. Schwartz' early experiments were single blind, so double and triple blind protocols were developed. Dr. Schwartz agreed on the need for these improvements and a great deal of time was spent developing and incorporating them into the study protocols. As Dr. Hyman must know, and as I previously pointed out, it is difficult to obtain funding for something so far from mainstream science. So I think it is understandable that Dr. Schwartz did not meet all of these criteria in his initial studies.

I am going to comment on a few of the points made by Dr. Hyman in his critique that apply directly to the readings in which I was involved.

1. As a young man, Dr. Hyman was involved in palmistry, and learned that people can easily be convinced, or can convince themselves, of the validity of things that he told them he learned from reading their palms. I have no argument with this. I have seen this sort of thing, and I believe we all are guilty of occasionally creating our own reality. The problem is that he implies that people who believe in paranormal phenomena automatically fall into this same category. This may be true in some cases, but *always*?

2. Like other critics, he suggests that the mediums are getting specific information about the sitter by employing a technique known as "cold reading". This is a technique used by many of the fraudulent practitioners of mediumship, who throw out

suggestions, for example, the first letter of a name, and observe the sitter's reaction. They repeat this until they get some indication, either spoken or through some sort of body language, that their statement is correct. If this doesn't work, they move on to something else. Dr. Hyman also refers a number of times to sensory leakage—the inadvertent leaking of information, through lack of proper controls, from the sitter, or possibly the researcher, to the medium.

Clearly, neither of these situations could have occurred in the readings I received and referenced in previous chapters. These readings were received **over the Internet**, from mediums hundreds of miles away, one day after I met Dr. Schwartz. There was absolutely no contact between me (or my wife) and the mediums. The mediums and I had never even heard of each other. So a skeptic would then have to fall back to the "he wants to believe so he is making the data fit" argument. To show how easy it is to make the data fit virtually anyone's situation, Dr. Hyman takes a few statements made by a medium in Dr. Schwartz' book and shows how this could be interpreted to fit his situation.

We all have many things in common with others—hair and eye color, food preferences, names, favorite sports, places of residence and so forth, so obviously it is not difficult to make statements intended for one person fit others. If I use the charades reading as an example, nearly everything in that reading could apply to others. For example, many people grow up in poverty, some wives die years before the husband, many people travel extensively, are religious, live in Hawaii and work in their garden, etc. The key point here is that all of these statements fit my dad. If the medium had stated that he came from wealth, or had died years before his wife, or had never traveled, or was an atheist, or had lived in Minnesota and went ice fishing—any of these would score as a miss. It doesn't matter how many others the statements fit as long as they fit my dad also.

From a scoring (and believability) standpoint I look at three factors:

- Are there a significant number of statements that fit the deceased, regardless of how many others they may fit? If so, then the reading gets a good score for that data. This is often data that helps identify the deceased.

- Are there what we sometimes call "dazzle shots"—statements that are truly unique and that do not apply to many others. The comment about my dad's throat is a "dazzle shot". That was the statement that made the reading even more believable for me. Other examples are the "bump-on-the-head" comment discussed in Chapters 8 & 9, or when Janet Mayer is making statements that apply to my son and suddenly switches to a young girl with comments about a starry night.

- Does the personality being revealed fit the deceased? In my experience, this is also usually a factor. My dad's personality was evident in the charades reading, and when I first noticed this it gave me a much stronger feeling of authenticity.

If all of these factors are present, we have what I would consider a reading that scores well, and one that is meaningful to the sitter. Considering how vague some of the information presented to the mediums can be, (as explained in Ch 9), I am impressed with how well many mediums are able to do. On the other hand, if there were more incorrect than correct statements, and no dazzle shots, the reading would not score well, and would probably result in a disappointed sitter.

3. Dr. Hyman states that the sitters were *deliberately* selected because they were already disposed toward

the survival hypothesis. I don't know how many of the people that were tested were predisposed, but I can assure you that I was not. I had no expectations, and was surprised at the number of hits in the readings I received. I was involved in a number of experiments, and I am not aware of anyone being rejected because they did not meet some "predisposition" criteria. If they were being screened in this way, I would not have qualified. Perhaps a future experiment can be done where only pure skeptics are tested. The question then becomes whether or not they would be objective in their evaluations. I would love to see funding for studies of this type.

4. Reliance on uncorroborated sitter ratings. Dr. Hyman states that "The 'accuracy' ratings of the mediums depend entirely upon the judgments of the individual sitters. Each sitter is solely responsible for validating the reading given to him or her." (Hyman, 2003, pg. 8).

It is not clear to me who Dr. Hyman thinks should validate the statements made by the medium, but I can think of no one better qualified than the sitter. I assume that Dr. Hyman suggests this as a precaution against fraud. If the sitter wants to lie or "stretch the truth" about the validity of statements made by the medium in a reading being done for them, they are only cheating themselves. I suppose in a perfect world, independent corroboration of every statement would be done, but doing so would require an inordinate amount of time, and therefore money. I don't think it is a practical expectation to be that rigorous, and in my opinion it is generally not necessary. He also suggests that Dr. Schwartz should have collected more data and followed it with studies conducted using generally accepted scientific criteria before publishing his results. Now that I am aware of the time and cost involved, I'm glad that I didn't have to wait the years it may have taken for this to be accomplished before Dr. Schwartz published his *Afterlife Experiments* book.

By publishing his book when he did, Dr. Schwartz was able to get others involved which then resulted in the more rapid development of improved methodologies.

There are fifteen pages to this article, so I will leave it to the reader to review it in more detail if they have further interest. This critique was written about ten years ago, and I think some of the points made by Dr. Hyman are useful in helping to refine study protocols in the future. However, I don't think he was being objective when stating that Dr. Schwartz was "badly mistaken", and he certainly didn't provide evidence to support his statement.

I recently read another critique about paranormal phenomena. It is an online debate published around March, 2010 by Michael Shermer in a publication entitled SKEPTIC. The debate is between Shermer and Deepak Chopra.

Dr. Shermer initially studied Christian theology as an undergraduate, but switched to psychology and has a bachelors degree in psychology/biology. He then received a masters degree in experimental psychology and later earned a PhD in the history of science. Since 2007, he has been an adjunct professor at Claremont Graduate University. (Wikipedia).

Deepak Chopra is an Indian medical doctor, public speaker, negotiator, and writer. Chopra began his career as an endocrinologist and later shifted his focus to alternative medicine. He now runs his own medical center with a focus on mind-body connections. He is also a lecturer at the Update in Internal Medicine event, sponsored by Harvard Medical School's Department of Continuing Education and the Department of Medicine. (Wikipedia).

The article begins by noting that Dr. Chopra and Dr. Shermer had a previous debate discussing the virtues and value of skepticism. Dr. Chopra wrote a book on the subject, *Life After Death: The Burden of Proof*, published in 2006.

This is a long debate so I am going to comment on items that are relevant to what I have been discussing. I will leave it to the reader to review the article in greater detail.

Dr. Shermer defines the soul as "the unique pattern of information that represents the essence of a person. By this definition, unless there is some medium to retain the pattern of our personal information after we die, our soul dies with us".

So far, we are in agreement. He goes on to say that:

"Our bodies are made of proteins, coded by our DNA, so with the disintegration of our DNA our protein patterns are lost forever. Our memories and personality are stored in the patterns of neurons firing in our brain, so when those neurons die, it spells the death of our memories and personality, similar to the ravages of stroke and Alzheimer's disease, only final."

(Shermer, 2010, pg. 2).

He then gives the reasons for his belief, which the reader may want to review.

One of my primary reasons for writing this book is to present the case that the mind (or soul) is separate from the brain, and that in fact it survives the death of the brain.

In Ch 12, I point out that our personal memories, which are distinguished from our physical (DNA) memories, may be either transferred from our brain at death, or were never stored there in the first place.

In Ch 13, Rupert Sheldrake is quoted as pointing out that, although they have tried, scientists have never been able to prove that our personal memories are stored in our brain. The question is whether or not the brain may function as an antenna to send and receive signals, which may then activate controlling functions within the brain. Damage to the brain will inhibit it's ability to function normally. This disability will affect the brain's control of bodily functions, but it may also affect normal functioning of the brain as the controlling interface in the transfer of information with sources external to the brain.

In Ch 14, I give my personal reasons for believing that we are made up of two basic "parts"—our physical body with it's brain, and a second part, our mind, that is defined by our feelings and emotions. This non-physical part is usually called "consciousness" by traditional scientists.

In the next chapter (Ch 16), I will explain why I am not able to accept the theory that life is the result of the gradual evolution of our DNA. Physical life may be—but not our minds.

In summary, I would say that Dr. Shermer and I are on opposite sides of the fence on this one, and it will probably remain a topic of debate within the scientific community until one of these views is proven to be correct.

Dr. Shermer goes on to address the six lines of evidence cited by Dr. Chopra in his book that convince him that the soul is real and eternal. Those six lines of evidence are: Near-Death Experiences (NDEs) and Altered States of Consciousness, ESP and Evidence of Mind, Quantum Consciousness, Psychic Mediumship and Talking to the Dead, Prayer and Healing Studies, and Information Fields, Morphic Resonance and The Universal Life Force.

Dr. Chopra does a far more thorough job of addressing Dr. Shermer's comments than I could, so I suggest that, if interested, the reader examine this material. I will comment on a few of Dr. Shermer's statements that are relevant to what I have discussed in previous chapters.

1. He states that new evidence shows that NDEs are, in fact, a product of the brain, and cites a study by Michael Persinger who induced NDE symptoms in subjects by subjecting their temporal lobes to patterns of magnetic fields. The problem is that his conclusion is based on the assumption that the magnetic field is affecting only the brain itself. He does not consider the possibility that there may be an external field associated with brain activity that is also being affected. In other words, he is simply stating the materialist's view that our memories and personalities are stored in the brain.

2. He discusses the effects that various chemicals have on the brain, and concludes that "NDEs and OBEs are forms of wild 'trips' induced by the extreme trauma of dying." (Shermer, 2010, pg. 6)

This is the same type of response that Melvin Morse received from his colleagues when describing his patient Katie's experience (see Ch 4). There is no question that ingesting hallucinatory drugs has an effect on the brain. It does not follow from this that everyone who has a Near Death or Out Of Body Experience is hallucinating, and I am not aware of any studies that support this opinion.

3. When discussing mediumship, he again brings up the subject of "cold reading", something I have already addressed. He also mentions something called warm reading which "utilizes known principles of psychology that apply to nearly everyone", and cites a book by Ian Rowland, a mentalist and magician, titled *"Full*

Facts Book of Cold Reading." This book "provides a list of high probability guesses, including identifying such items as found in most homes that are sure to convince the mark that their loved one is in the room." (Shermer, 2010, pgs. 10-11).

It sounds to me like a how-to-do-it book for people who want to become proficient at fraudulent mediumship, as implied by the word "mark." These techniques are also used by those involved in what is sometimes called "psychic entertainment."

There definitely are fraudulent mediums and there is no doubt that many people are vulnerable to the techniques used by these people. However, as I have previously stated, I have been involved closely with mediumship studies and they were designed and conducted in a way to rule out fraud. It is very misleading for skeptics such as Hyman and Shermer to imply that fraud is usually employed when practicing mediumship.

There is a good deal of "food for thought" in this debate, so I recommend it for those who are interested. If you decide to read the main article, I think it would be both interesting and informative if you also read the comments, both pro and con, submitted by those who have read the debate. You can tell from some of the comments those people who think only in terms of the physical world, and those who have a much broader view of life.

When criticizing the work done by paranormal investigators, skeptics sometimes cite a lack of peer review. I'm sure this is a valid criticism in some cases. However, I do know that research organizations such as the Windbridge Institute and the Institute of Noetic Sciences publish their work in peer reviewed journals. A listing of some of these publications are provided on the WBI website.

The number of investigators involved in the scientific study of paranormal phenomena is increasing. I have mentioned a few of them throughout this book, and referenced several more at the end of the book in Recommended Reading. These people are continuing to advance our understanding of these unusual phenomena. It is clear that we could use more legitimate scientists to challenge the work all of these people are doing, to try to replicate the results being obtained and to suggest better ways to test their hypotheses and to improve the tests that are being done.

Many of the scientists conducting studies in the area of paranormal phenomena were generally quite skeptical when they first became involved, but when the results began to show evidence supporting the survival-of-consciousness hypothesis, their skepticism was replaced by a more objective attitude when looking at the data. I think it can be said that we need objective scientists—and we have them. In general, those engaged in scientific investigations are publishing their results in peer-reviewed journals and books. As I see it, these people are the real skeptics. As the definition states: "Philosophical skepticism is an overall approach that requires all information to be well supported by evidence." As far as I'm concerned, this should be required of those on both sides of any debate. The people doing experimental studies are gradually providing an increasingly large amount of evidence to support the hypothesis that an afterlife exists.

Rupert Sheldrake's newest book, entitled *Science Set Free*, was published in the USA in Sept. 2012. In the Introduction, he points out that the "scientific worldview" is immensely influential because the sciences have been so successful. However, he then lists ten core beliefs of modern science and shows the ways in which science is being held back by centuries-old assumptions that have hardened into dogmas. In addition to exposing these dogmas, Sheldrake provides a detailed history of how they evolved from the beginning of the modern scientific era. This book will provide plenty of food for thought for anyone who is curious about how we arrived at our current position of materialism.

Another excellent book is *The Synchronized Universe* by Claude Swanson, Ph D. As related on the book's cover:

"Dr. Claude Swanson was educated as a physicist at M.I.T. and Princeton University. He wanted to understand the Universe at the deepest level. Then one day he discovered that his scientific education had left out a few things—

The parts left out make up the new scientific revolution. Modern physics has suppressed and ignored the paranormal, but in the laboratory paranormal phenomena are now a proven fact and are beginning to shake the very foundation of physics. The new scientific revolution is changing our understanding of the universe and of ourselves—

CHAPTER 16

Science and Spirituality

In three words I can sum up everything I've learned about life; it goes on.

Robert Frost

It is interesting how, at times, things seem to happen serendipitously. As I neared the end of writing the previous chapter, I was thinking about the similarities between the information in some of the books on afterlife phenomena and, in general, what is taught by traditional religions. At about the same time, without specifically searching for them, I found out about two recently published books whose authors have spent a large part of their lives researching afterlife phenomena. Both authors have a Christian background, but approach the subject from a non-fundamentalist viewpoint. Based on their analysis of the material contained in many books on subjects such as mediumship and NDEs, these authors present a picture of what they believe the afterlife may be like.

When I first glanced at these two books, my natural skepticism, combined with my having rejected traditional religious teachings, caused all sorts of warning bells to go off in my head. This reaction always causes me to be more critical than usual when reading books that appear to be far from mainstream beliefs. However, because of this tendency toward a bias, I feel it is especially important to keep an open mind. I decided to read these books because much of the material was obtained by analyzing articles on mediumship going back more than 150 years, and mediumship is an area of study that has gained a great deal of credibility with me.

The fact that we are not yet able to scientifically prove beyond *any* doubt what is stated about a hereafter does not make the information false. In fact, many of the concepts expressed in these books seem more reasonable to me than the origin-of-life theories I have heard proposed by materialistic scientists. For purposes of clarification, I am going to reiterate what I mean by "materialistic". Wikipedia states it well: "In philosophy, the theory of materialism holds that the only thing that exists is matter or energy; that all things are composed of material and all phenomena (including consciousness) are the result of material interactions. In other words, matter is the only substance, and reality is identical with the actually occurring states of energy and matter."

When I think about materialistic theories of the origin-of-life, I think of evolution. I am not well-versed on theories of evolution, but I have read and tried to understand what I believe is the most popular theory. From what I can find, this theory states that life evolved some 3 billion years ago (following the formation of the Earth) from some sort of primordial chemical soup. The process began with basic building blocks such as simple amino acids and sugars and gradually evolved into more complex structures such as proteins and DNA. From there the process, guided somehow by RNA, eventually resulted in the complex, presumably intelligent organisms we call humans. I couldn't find out how the controlling element, RNA, was formed. If I sound confused, it's because I am. Every time I try to read and understand these theories I get more confused.

From what I can gather, they are an extension of the findings of Charles Darwin. Darwin's observations show that there are continual changes to physical life on Earth, and we now know that this is also true of the cosmos. In my opinion, too many people assume that it will eventually lead to an explanation of everything about our existence, including the beginnings of life.

I think it is reasonable to expect that someday theories of evolution may lead to an explanation of the formation and evolution of physical (material) life, including our bodies, but I would also expect there to be some sort of defined controlling input as part of the process. As I stated in Chapter 14, I don't believe that our physical bodies alone define who we are. The concept of dualism makes more sense to me. Because I find the materialistic theories to be a really big stretch, I like to read books that approach the subject from other viewpoints.

My main purpose in presenting these books is to give the reader an overview of what others see as the afterlife. I am not presenting the material in order to convince anyone of it's validity, but rather to present views other than those of mainstream science and mainstream religion. The material is probably easy for many religious people to believe and a materialistic skeptic would most likely find it very difficult, if not impossible, to accept.

The first of these books is *The Fun of Dying* by Roberta Grimes published in 2010. Roberta is an attorney who "had an experience of light at the age of eight and spent the next half-century figuring it out". (Grimes, 2010, pg. 1).

In the introduction, she points out that we have been collecting data related to the death experience and what follows it for a century and a half, but states that "—to talk about dying with authority is to set oneself against those two modern bastions, mainstream science and mainstream religion, which long ago carved up all truth between them and thereby have shut out any facts that neither of them wants to own." (Grimes, 2010, pg. 2).

She further states: "Mainstream scientists believe that there is nothing after death to discover, while mainstream clergymen are sure that there is nothing more to know." (Grimes, 2010, pg. 7).

I think she expresses the current situation, at least in our culture, quite well.

Grimes makes some statements that at first may seem difficult to grasp. However, if you think of them in the context of what has been learned in the last few decades about the physical world and emerging evidence supporting the existence of a metaphysical world, they make more sense.

In discussing mainstream science vs. mainstream religion, she makes the following statements. Regarding mainstream science:

- "Physics is the core science on which the other scientific disciplines are based, and until about a hundred years ago it made steady and impressive progress. Then early in the 20th century physics more or less ran off the rails when it's two fundamental concepts—Newtonian physics, which works on large objects; and quantum physics, which works at the subatomic level—were found to be incompatible. To vastly simplify what is a big problem, since both Newtonian and quantum physics are well-proven theories in their own domains, physicists have spent the past century searching with an ever-increasing sense of frustration for a theory that encompasses them both. But perhaps such a unified theory won't be found, because (as you will shortly see) the very matter and energy and time and space that physicists are trying to understand do not exist in most of reality." (Grimes, 2010, pgs. 7-8).

I found that last statement rather intriguing, and at first somewhat difficult to understand. For those who may have an interest, I will give a summary of the point she is making. She states that:

"Quantum physics is a theory that posits that subatomic units called quanta are the building blocks of reality. Matter, energy, time and space are all composed of some variant of quanta, and material quanta appear to be both particles of matter and waves of energy."

"Here are two facts derived from the quantum physics literature which help us begin to understand that matter, energy, time and space may not be objectively real:

1. Things don't achieve a fixed reality until they are consciously observed.

2. Measuring a quality of one subatomic particle instantly affects the same quality of a synchronized particle, even if the particles are widely separated."

"The following fact also seizes my imagination, but it is probably a coincidence:

3. Most of the universe is invisible to us. Physicists' understanding of gravity and other forces suggests that as much as 96% of the universe is composed of "dark" energy and matter that is invisible to us, and for now, inexplicable." (Grimes, 2010, pgs. 28-29).

Further attempts to explain these complex theories are well beyond the scope of this book.

Regarding mainstream religion, she points out that:

- "Many scientists believe in God, but mainstream science's dogmas and protocols have long been fundamentally atheistic. In that, modern science has become a belief-system not unlike a religion."

- "Religions are belief-systems. They are based upon faith and not upon facts; or, more precisely, they are based upon modern faith in a set of ancient facts."

- "Some of the Eastern religions have for millennia held theories about death which have turned out to be pretty close to what evidence now tells us is true." (Grimes, 2010, pgs. 8-9).

Further on in this chapter, I summarize the religious beliefs of five religions, including Hinduism and Buddhism.

In general, I agree with her comments. She suggests that we need to open our minds, and instead of separating material and spiritual truths into two competing camps, we should use both scientific and religious facts to validate what afterlife evidence now tells us. In order to ascertain what she feels are spiritual truths, she decided to read only the Gospels, and to ignore the interpretations placed on them by others, such as the apostles and their followers. This was a revelation for her, and allowed her to reconcile the conflicts she felt when reading the complete Bible. She shows how modern evidence of the afterlife is supported by what is revealed in the Gospels.

In Chapter 8 Grimes provides a listing of the major elements of the afterlife (what she calls "Summerland") as revealed by discarnates through mediums. These statements, compiled from the various literature that she read, are listed below. I have shortened some, and not shown all of them.

- To answer a common question first, some people do attend their own funerals. Not everyone does it, and many attendees will have helpers there to comfort them and draw them away if necessary.

- Few dead people care about their former bodies. Some express frustration that their loved ones are wasting energy and emotion in visiting and tending what they consider to be empty graves.

- Dying does not in itself bring enlightenment. This often surprises the recently dead, who express frustration at finding themselves exactly the same people that they were before they died.

- We live on the middle post-death levels in a constant pure white light that is brighter and more diffuse than sunlight, although it doesn't affect our eyes in the way that sunlight would.

- There is no sun, there are no stars, there is no night, and it never snows or rains, but if you want to enjoy a sunset or night or snow or rain you can have them.

- Breathing here fills our bodies with a living and energizing air which is all the nourishment we need.

- Although we don't need to eat or drink, if we want to eat or drink we can have what we like and enjoy feasting for as long as we like. We lack digestive organs, so our illusory food and drink disappears.

- After the post-death nap that most people take, we never need to sleep again.

- We can appear at whatever age we prefer. Most people choose to be the person they were in their prime of life, but that is up to them.

- The after-death levels are a spiritual hierarchy in which we all are working to advance, but without objective time there is no sense of hurry to our progress.

- Here our earth-status counts for nothing.

- Most after-death communication is telepathic. Somehow this does away with the problem of different earth-languages, so once we are accustomed to telepathy, we can converse with anyone we like.

- We can read the minds of those we left on earth, and we can communicate with them by thought, but living people seldom notice our messages. A few of the dead are able to manifest on earth physically under certain conditions, and some can affect electricity, but most of us are unable to do more than produce a whiff of some distinctive scent.

- Colors in the Summerland are not limited to the visible-light spectrum, so there are colors here that are impossible for the living to imagine.

- Flowers are everywhere in the Summerland.

- Water is abundant here. It is crystalline and it sparkles like earth-water, but it is very different indeed." (Grimes, 2010, pgs. 72-75).

There are a few more pages of these descriptions. I have listed the first few pages to give the reader an idea of the books overall content. Appendix I contains references to many books from which she has obtained her material. Appendix II, titled "Listening to Yeshua" (the Aramaic name for Jesus) contains quotes from Christian scriptures. She states: "Yeshua's words are amazingly consistent with afterlife-related evidence, but a lot of what mainstream Christianity teaches is not." (Grimes, 2010, pg 127).

If you are interested in reading an uplifting, encouraging book about what may come after we leave this "life", try this book. For some, especially non-religious skeptics, it will require an open mind to accept what she says as at least plausible. The author's writing style makes it an easy book to read.

The second book, published in 2011, is *The Afterlife Unveiled: What the Dead are Telling Us About Their World* by Stafford Betty, who is a Professor of Religion at California State University in Bakersfield, and is a world expert on afterlife studies.

Betty provides seven accounts of the afterlife "allegedly conveyed by spirits who are there" through seven different mediums. Some of these spirits 'died' centuries ago, others more recently. In the Conclusion, he lists specific details of the afterlife revealed by these spirits, some of which I will list here. Again, I have shortened some of these descriptions, and I have not listed all of them.

- Our present ideas about heaven and hell are illusory. Hell is hellish, but it's not a place of physical pain, nor is it a place where there is not help. And heaven is not one place but a spectrum of worlds stretching from the lowly joys of souls newly arrived to spheres of unimaginable bliss and perfection for souls far more advanced. The Afterworld is not some fantastic vision of infinity where souls are locked in poses of permanent rapture gazing at the face of God. And no one floats on a cloud while playing a harp. Rather it is a place with landscapes and seas and houses and cities reminiscent of our own world—a material world, but of higher vibrations insensible to us earthlings. There are gardens, universities, libraries and hospices for the newly dead—but no factories, fire stations, sanitary landfills or smokestacks. There are no dirty jobs to do.

- The Afterworld begins at the earth's surface and extends outward. Earth is the nucleus of the entire world system that the spirits describe. Many spirit communicators tell us that their world "envelopes and interpenetrates the physical world."

- Spirit realms vary from culture to culture. We should not expect the Eskimo's afterworld to look like the Maori's. The laws governing their worlds will be the same, but the appearances will vary.

- Earth's slow "vibrations"—we don't have a scientific understanding of what this word means, but it turns up everywhere in spirit communications—dumb down our ability to sense the presence of spirit, including the Divine. A quickened vibration, such as we find in the afterworld, or what we call the astral, greatly increases one's sensitivity to spirit. The Divine is no closer to the astral world than to our own, but spirits can discern or intuit It more cleanly.

- The newly 'dead' are thoroughly themselves when they pass. Their personalities and habits and character, for better or worse, are completely intact. Nothing miraculous happens to them when they pass. Their astral body is not a 'resurrected' body, but was always present as the soul's 'inner envelope' while embodied in earthly flesh. Once the physical body dies, the inner body quite naturally becomes the outer—as a snake's inner skin becomes the outer skin once it sheds the old. There is nothing miraculous about the process of surviving death.

- So natural is the process of dying that many souls do not realize at first they have died. The difference in appearance between the physical and the astral body is relatively slight; so is the difference, as we've seen, between the new world they've entered and the old one left behind.

- Spirits are not omniscient. They don't get the answers to all the questions that puzzled them back on earth just because they've "died".

- Though they are recognizably themselves, life in the higher astral—a term commonly used by spirits to describe the light matter of their worlds—is more vivid and intense than on earth, not more ghostly. They often describe their world as more real than ours. Our earth is the copy or facsimile of theirs.

- Astral beings have fewer limitations. They can communicate telepathically and with much greater precision than through the cumbersome medium of speech. They can move from place to place by willing to be at their destination, though they can walk if they want to. Their minds are sharper, their emotions more acutely felt, both positive and negative. They see and hear as before, but in a more intense way.

- Some spirits describe a phenomenon known as the Akashic or etheric records.

- Because experience is heightened, pain as well as pleasure is intensified, and sometimes the pain is acute. It comes from an awkward awareness of all the pain inflicted on others by one's cruel or insensitive actions or words, which are now experienced as one's own.

- The old or decrepit or injured bodies they left behind do not follow them beyond death. Yet they are embodied, almost always in a manner that makes them recognizable by others who knew them on earth. At all times they bear with them an astral signature that identifies who they are.

- Spirits greatly respect time spent on earth.

- The Creator places souls in the difficult environment of earth because He (She, It) loves them. He wants to see them grow in wisdom, love and power. He knows that the only way to bring out the best in a soul is to challenge it, in the same way that a good teacher challenges her students.

- The Afterworld is a broad-based society of every conceivable kind of person, most of them flawed and incomplete in some way or another. Many are no more motivated to 'grow their souls' than they were back on earth. But many are determined to advance and they do so.

- The astral world provides opportunities for every wholesome interest or avocation—from science to music to theology to astral architecture to homebuilding. It is a joyful, endlessly fascinating place, full of challenges, for those mature enough to value it.

- Physical danger does not exist in the astral. Neither does physical illness. Eating is optional and sleep unnecessary.

- Many astral inhabitants maintain a lively interest in the events of earth and long to help it progress. They claim that many or even most of earth's most brilliant achievements were inspired by spirits telepathically projecting their ideas.

- Spirits do not forget their loved ones back on earth, whom they often seek to help by projecting healing thoughts or by personal visits. Some spirits enjoy sitting unseen next to their loved ones on earth, whom they miss just as we miss each other. Others try to communicate with loved ones on earth through mediums.

- There are hellish regions in the astral, and large populations that make their home there. What is sometimes referred to as the Shadowlands is a vast world of many conditions. The landscapes vary from sordid city neighborhoods to parched, gray scrubland to dark, lifeless deserts. The vivid clarity of higher realms is missing. Instead there is a dull overcast. Temporarily lost or confused or stubbornly unrepentant souls populate these regions.

- The worst of these souls aggressively seek to harm vulnerable humans on earth. They band together to resist progress and truth—.

- Missionary spirits minister to souls in the Shadowlands. Residents can free themselves if they are willing to face up humbly to their errors and crimes and repent them. Some do; and most, perhaps all, will eventually. No spirit is condemned forever to the dark regions.

- There are no rigid creeds or magical beliefs that souls have to accept. Whether you are a Baptist or a Catholic or a Mormon or a Hindu or a Buddhist or a Muslim or an Anglican is of no importance. Many of earth's favorite religious dogmas are off the mark anyway, and the sooner they are recognized as such, the better.

- There are no masks in the astral. You cannot hide from others what you are; the quality of light shining forth from your body tells all.

- The sense of earth time fades fast. There is duration, but nothing like clock-time with its schedules and deadlines.

- Spouses, relatives, friends, and former teachers, some from earlier lives, some long forgotten, turn up and may renew old friendships. (Betty, 2011, pgs. 102-107).

Betty also discusses the subject of reincarnation, and concludes that it is a real phenomenon. He states: "For me, AD Mattson (one of the spirit communicators in his book) sums up my belief best: You can elect not to return, and many do, after they have achieved a certain spiritual development. But the physical plane is a 'school' for learning and development, and so most souls do desire to return for a series of incarnations." (Betty, 2011, pg. 111).

As with Grimes book, I think those interested in a vision of what the afterlife may be like would also find this an interesting, well-written enlightening and easy-to-read book. Both authors present a picture suggesting that the afterlife is similar in many respects to life on earth, in contrast to the concept presented by some religions of pure bliss for believers and condemnation to hell for eternity if you are a non-believer or sinner.

There is quite a bit of overlap in the material presented in these books. A few of the items are also supported by my own observations and experiences. Some examples:

1. Grimes states: "We can appear at whatever age we prefer."

In Ch.17, I discuss lucid dreams that I have had in which John at times appears to be very young, and at other times to be in his teens or early twenties. However, there is never any doubt in my mind while in these dreams that it is actually John. Also, Janet Mayer received an impression that indicated a young boy in the first reading about John. (see Ch 6).

2. From Betty: "The newly 'dead' are thoroughly themselves when they pass. Their personalities and habits and character, for better or worse, are completely intact."

And Grimes states:—the recently dead, who express frustration at finding themselves exactly the same people that they were before they died."

As I mentioned in the charades article, the thing that gave that reading a much higher degree of credibility with me is that the personalities of the individuals, such as my dad, were so evident.

3. From Grimes: "Most after-death communication is telepathic. We can read the minds of those we left on earth, and we can communicate with them by thought."

And from Betty: "They can communicate telepathically and with much greater precision than through the cumbersome medium of speech.

They claim that many or even most of earth's most brilliant achievements were inspired by spirits telepathically projecting their ideas."

Telepathy is defined as the transmission of information from one person to another without using any of our known sensory channels or physical interaction. The concepts put forth by Sheldrake and Baumann in Ch. 13 suggest a form of telepathy. Studies supporting the validity of the concept of telepathy have been conducted and appear in the literature, such as those done by Dr. Dean Radin and Dr. Daryl Bem. I reference a website at the back of this book where this information is listed.

Sometimes I ask myself if I am now finding so much of this material believable because it makes me feel better, or because I need an answer, or something of the sort. The answer that keeps coming back is this; if what I learned from my own experience with mediums is believable, then why would I not believe what has been written by other reputable people on the subject? If I am able to look at this information without my cultural, scientific bias, I find it easier to accept. Of course, nothing is certain, but I think a much stronger case exists for the concept of survival-of-consciousness than for the "you're dead, it's over" concept.

What gives me a sense of closure is that I believe I have found many of the answers I swore I would search for the night I lost my son. I began to get answers in both the initial readings I received and in our Asking Questions experiments, using the same method as related in these two books—mediums. What we have done that gives credibility to the reports in Grimes', Betty's and other books on this subject is experiments using as many tenets of the scientific method as possible, given the phenomenon we are studying. Can I say that these are the final answers? No, I think we've got a long way to go in terms of scientific proof, but I am encouraged by the results to date. Another of my goals in writing this book is to make people aware of literature that provides views of our existence that differ from mainstream religious and scientific views.

I was on a flight while finishing *The Afterlife Unveiled* and ended up talking to the man sitting next to me, Chris, about the book I was attempting to write. Chris is a practicing Christian. After listening to me share what I've learned from NDE books and mediums, he asked me how I was going to lead the reader from the point where I show that there is an afterlife to the point where they meet God—in his case, Christ. Our conversation made me realize that many religious individual's goal is to follow the teachings of their religion so that they can be with God after they die. There are certain things that you should do, and other things that you must not do, if you want to get into Heaven. My conversation with Chris made me realize that my goal is not to try to be with God when I die. I just want to see my sons, my mom and dad and all the others that I'm connected to.

The other thing Chris got me thinking about is the various religious teachings. I am not well versed on the worlds religions, so I turned to Google, where I found a website called "Religious Tolerance". I read their descriptions of what I think of as the five main religions: Buddhism, Christianity, Hinduism, Islam and Judaism, and discovered some interesting facts, some of which surprised me. I have paraphrased some of the material.

1. <u>Christianity</u> is the worlds largest religion (not a surprise) and about 33% of the worlds population consider themselves Christian. About half are Roman Catholic. There are on the order of 1500 different Christian faith groups (a big surprise)—which promote many different and conflicting beliefs (not a surprise).

2. <u>Islam</u> is the world's second largest religion, about 23% of the world's population, but growing rapidly and if current trends continue, Islam will become the most popular world religion in the mid-21st century.

If the reader is interested, the website has a chart comparing Christian and Islam beliefs. There are several similarities and also notable differences.

3. Hinduism is the world's third largest religion, and is generally regarded as the world's oldest organized religion. It consists of thousands of different religious groups that have evolved in India since 1500 BCE. Because of the wide variety of Hindu traditions, freedom of belief and practice are notable features of Hinduism. Most forms of Hinduism are henotheistic religions. They recognize a single deity, and view other Gods and Goddesses as manifestations or aspects of that supreme God.

4. Buddhism, the world's fourth largest religion, is divided into a number of different traditions. However, most traditions share a common set of fundamental beliefs. One is reincarnation—the concept that people are reborn after dying.—A practicing Buddhist differentiates between the concepts of rebirth and reincarnation. In reincarnation, the individual may recur repeatedly. In rebirth, a person does not necessarily return to earth as the same entity ever again.

The website contains a discussion about whether Buddhism is a religion or a philosophy.

5. Judaism. There are five main forms of Judaism in the world today.

- Conservative Judaism began as a reaction against the Reform movement. It is a mainline movement midway between Reform and Orthodox.

- Humanistic Judaism is a very small group, mainly composed of atheists and agnostics, who regard mankind as the measure of all things.

- Orthodox Judaism is the oldest, most conservative, and most diverse form of Judaism.—They attempt to follow the original form of Judaism as they view it to be.

- Reconstructionist Judaism is a new, small, liberal movement.—They reject the concept that Jews are a uniquely favored and chosen people.

- Reform Judaism is a liberal group followed by many North American Jews. They follow the ethical laws of Judaism, but leave up to the individual the decision whether to follow or ignore the dietary and other traditional laws.

Regarding what Jews believe, the website states: "This is a far more difficult question than you might expect. Judaism has no dogma, no formal set of beliefs that one must hold to be a Jew. In Judaism, actions are far more important than beliefs, although there is certainly a place for belief within Judaism."

A book that gives an excellent overview of the teachings of the worlds major religions regarding an afterlife is *Handbook To The Afterlife* by Pamela Rae Heath and Jon Klimo.

The majority of the earth's population believe in an afterlife. They differ in many of the details, but a large part of this can be attributed to many individuals who have added their biases. I often wonder what makes the concept of an afterlife so difficult to accept within the mainstream scientific community. It's apparently the need for everything to be proven using accepted scientific principles, so I am encouraged that we are now making great strides forward in this area of science.

I too now believe in an afterlife. The readings I received were just too convincing for me to have much doubt. The twelve years of religious schooling I had did not present much of an image, but the books by Grimes and Betty, and others I have read, have at least given me an indication of what that afterlife may be like.

The image I now have is that of an existence where we still have a body, but one that is ethereal (see Glossary). We are able to travel effortlessly, unhindered by a heavy physical body. We can communicate with others without the difficulties of language barriers or miscommunications. Our world is still material, and in many ways similar to our earthly world, but it too is ethereal as well as being a more vivid, colorful and pleasant place. The messages I received from my son, my dad and others give it the appearance of being a cheerful place. There is plenty to do, including working toward a higher level of existence if we want. We are able to be with those we care about who have also passed. Life is a continuum. There is no death that signals the end of our existence, only a transition from one state to another. We are able to observe, if we want, those still in earthly physical form, and—with the help of mediums—we can communicate with them.

At this point, this image can only be considered a belief, one that I'm sure will go through adjustments as we continue to learn more through advances in paranormal science. I'm in no hurry to find out how close I am to reality, but I'll be ready to go when my time comes.

CHAPTER 17

Letting Us Know They Are Around

> *It is through science that we prove, but through intuition that we discover.*
>
> *Henri Poincare*[7]

I now feel a welcome sort of freedom when reading books with content that is outside of mainstream subject material. I no longer have something nagging at me somewhere in the back of my mind, warning me not to believe the material. I have the feeling that it's OK to read it and then give it some thought, provided I maintain reasonable objectivity. Objectivity is a good thing, but to me an equally good and important thing is keeping an open mind.

The topics I discuss in this chapter, because they involve personal feelings and experiences, fall in the category of being subjective. However, I believe that, because these feelings are unusual and seem real to those experiencing them, they are deserving of consideration. The first topic is

---

[7] Jules Henri Poincare (April 29, 1854-July 17, 1912) was a French mathematician, theoretical physicist, engineer and a philosopher of science.

that of personally sensing, in some way, the presence of a deceased loved one.

Diane describes what she experienced the night that John was killed.

> "Much later that evening I sat quietly with my new grief, thinking in my mind, "where are you John?" John came to me and very simply said, "It's okay mom. It's no big deal." I felt a soothing, wonderful and definitely physical warmth . . . peace . . . love . . . . and actually joy. Only then did I sleep.
>
> I remember thinking that it was interesting that John came through to me in this way, because for some reason, I always believed that after a person had passed, if they came to you it most likely would be as an "apparition" at the foot of the bed, or somewhere up near the ceiling of the room and that it would be some type of "vision." This was not the case, for I felt John's presence deep within me—within my entire physical body. And it was definitely his voice, although I "heard" it telepathically . . . . not out loud. What he said to me certainly were his words! There's no way in my devastation that I would have said to myself, "it's no big deal." When John was here we had many interesting and philosophical discussions regarding the "meaning of life" and "where do we go from here?" In fact, in our last meeting shortly before his death we were having lunch together and that very discussion took place. Therefore, John's message to me: "It's ok mom, it's no big deal" made total sense to me, especially coming from a 22 year old! Short and sweet. I believe John was telling me that he found the answer . . . . at least in his own personal experience of death."

I remember, at the memorial service for my dad in Hawaii, going into the church, sitting in the front pew with the casket containing his body in front of me, and suddenly getting this very warm feeling. It made me feel so good that I began smiling, even though most people around me were looking very somber, and I remember thinking "that's not my dad in that box—that's just his worn out body, and he's letting me know it." This was several months before John died, and I think it was the first time I began to believe that perhaps we really do survive physical death.

My first memory is seeing my mom in a casket at the home of one of my aunts, and my dad telling me to "go over and tell your mom goodbye." That image is still very vivid in my mind. What I remember most is not only that I was not upset, but that I actually had this "good feeling" as I said goodbye. Sure, as a three year old the concept of death probably did not have the same meaning to me as it now has, and because my mom was isolated in the sanitarium I had not been able to get close to her, but that doesn't explain the "good feeling."

To me these may be subtle instances of the deceased letting us know they still exist.

Lucid Dreams

My definition of a Lucid Dream is a dream where you are aware, *while in the dream*, that you are having a dream. I had never had a dream like this until after John died. Much has been written about lucid dreams, but I am going to relate a couple that have occurred in my family. My wife Diane has always been spiritual (as opposed to religious) and like most people who are spiritual, she can accept most aspects of paranormal phenomena without having to question and analyze them the way I do. The following is a description she wrote of a dream she had shortly after John's death.

Dream of May 19, 1991

> "I asked where John was and was told that he was at a picnic, in a beautiful wooded setting. I was about to go look for him when he drove up on his motorcycle, telling me what a great time he was having, working there, socializing. He was filled with enthusiasm. He said there are "lots of sorority girls there!" He was SO HAPPY! He told me he needed to get back and why didn't I join him. YES, I wanted to! So I got onto my motorcycle(!) (which was all rusted and beat up), and just couldn't get it started. John had zoomed on ahead, figuring that I would catch up. Of course I was not able to join him . . . . yet.
>
> This is the first time I recall ever having what is called a "lucid" dream, where I truly felt I was receiving a visit from John."

Once in a while, Diane or one of my daughters would tell me how John had come to them in their dreams. I have to admit, I was happy for them but also a little jealous. What about me, John?? Then, about six months after his death, I was in Germany on a business trip and I was staying at my friend's offices and sleeping in a small loft he had there. One night while asleep I saw John walking with my dad! Yes, it was a dream but there was a key difference. I was aware that I was asleep and dreaming, and I had the very clear feeling that the three of us were together. After a minute or so, I woke up. My heart was pounding so hard I thought I was going to be bounced right out of the bed. This type of thing had never happened before. Finally, I was with John in a dream—and I knew it!

As Diane pointed out to me, probably the reason John had not been coming to me prior to that dream is because I was not receptive to the idea. I had to question and doubt everything. That first lucid dream gave me a tremendous amount of hope that I would definitely see my son again, and for several years after that I continued to have these dreams, most of which I have recorded in my notebooks. However, I did not have the "solid proof" that I seemed to need. I guess it was necessary for me to go through all of the reading and experimenting before I was able to finally feel that "now I know" I'm going to be with my son again, as well as with all of the other loved ones who have preceded me to the "other side."

The frequency of these lucid dreams has diminished now, and it may be because John knows that I finally feel certain that I will be with him again. Actually, from what I've learned in recent years, I strongly believe that we are together now—I just can't see or talk to him the way I'd like.

So it appears that in addition to the capabilities I previously listed at the end of Ch 14, another discarnate capability may be that they can appear to us when we are sleeping. I don't know what causes ordinary dreams, let alone lucid dreams, but the more I read and learn about paranormal phenomena, the more I am inclined to believe that our brains are, at least periodically, being influenced by outside sources. I am intrigued by the fact that, no matter what age or appearance John took in my lucid dreams, I had no doubt it was him.

From a scientific standpoint, one thing that seems possible to me is that discarnates are able to "telepathically" send a signal that contains an identifying code, similar to the "unique pattern of information that represents the essence of a person" described by Michael Shermer as existing in our brains prior to our death (see Ch. 15). The difference is that my hypothetical signals originate outside our brain and would be analogous to the coded signals used in transmitting radio, TV, cell phone or other electromagnetic signals such as those used to control spacecraft and to transmit images from satellites thousands of miles from Earth. This type of signal could also be the method, again hypothetically, used to transmit images, sounds and/or feelings to mediums (who may be more sensitive to these signals) or to the dying in deathbed visions. As I previously stated, I think this would be an interesting experimental subject to study.

Another theory is that, when dreaming, we are having an Out of Body Experience (OBE). Robert Monroe makes this point in his book *Journeys Out of the Body*. (Monroe, 1971, pgs. 203-204).

In this case, rather than signals being sent to our brain, what is considered our essence leaves our body, as described by people who have had a Near Death Experience. One difference between these two phenomena seems to be that those who experience an NDE appear to travel to another realm (the "other side") whereas those experiencing an OBE, from what I have learned so far, seem to remain "earthbound." In other words, an OBE appears to be only one element of an NDE. Logic leads me to believe that if people like Monroe could leave their bodies and travel to the other side they would do so. Apparently this only happens to those who are actually near death and they have no control over the process.

Janet Mayer was in town recently to give talks at a few local venues. As a result, I was finally able to meet Janet in person after ten years. The day she arrived, Janet, my wife Diane and I were seated in my house talking and waiting to leave for dinner. As we talked, Janet casually said something like "I'm getting an image of John with an older woman who has short hair. Now I'm being shown the color orange." As soon as she commented about the color orange, Diane immediately responded with "Ma!". Ma died several years ago and is the mother of a friend. She is the person Diane witnessed seeing "spirits" just before she died. Ma's daughter is a rabid San Francisco Giants baseball fan and is always wearing orange, the Giants team color, sometimes including her hair. A little while later when we went to dinner there was Ma's daughter, wearing an orange blouse, her fingernails painted orange, with her husband and another couple having dinner at the restaurant. Coincidence? Perhaps, but why would Janet, who had never even heard of Ma or her daughter, suddenly get those images?

Before ending this chapter, I'm going to tell of an incident that happened to me and that I sent to the Forever Family Foundation Signs of Life Newsletter. It falls in the category of Electronic Voice Phenomena (EVP) but I'm putting this one in just for fun.

Saved by Ricky!

I gave up on praying when I gave up on the Catholic religion as a youngster, and have never formally returned, except in cases of dire need. :-)

A few days ago I was pulled over by a police officer who said I had not made a complete stop at a stop sign. I was sure that I had, but she said it was not a "complete" stop. She then told me that they were required to take a video of all stops for use in court, and she went back to her car to check her video. If she wrote me a ticket, it would be around $200-$250 in California, points on your record, time in court or take an all-day class, etc.

Turns out she had pulled me over in front of the cemetery where my son Ricky is buried, so I began talking to Ricky, asking him to "do something—help me out here Ricky." I watched in my rear view mirror as the officer checked her video. Finally, she grabbed a piece of paper and walked back to my car. I figured I had a ticket, but she says "I'm going to have to let you go, sir. This has never happened to me before, but something is wrong with my video playback" and she let me go.

A skeptic would say that she went back and saw that I had made a complete stop, or that there was just a normal breakdown of some sort of her machine.

I know better! :-) I thanked Ricky profusely and went on my way.

Bill Kaspari
Summer, 2011

CHAPTER 18

The Pendulum Begins to Swing Back

> *Never doubt that a small group of thoughtful, committed citizens can change the world; indeed, it's the only thing that ever has.*
>
> *Margaret Mead*

I think most people probably wonder, at some point, what life is all about. As a youngster in Catholic school, I was told that our purpose in life is to know, love and serve God. I didn't really understand the concept of God, let alone why we had to be so "reverent." In our religion classes we spent a good deal of time reading about the history leading up to and during the time of Christ, as related in the Bible, but we also spent a great deal of time learning the rules, and the dire consequences that you would suffer if you didn't follow them. The impression I had was that I'd better be good and worship a certain way—or else. There was virtually no time spent reading the Bible. In fairness to my teachers, I know that they did their best to teach me right from wrong, as well as to prepare me for a higher education. I can't say what their perception of the afterlife might have been, but I doubt if it was the same as the one I now have. They didn't really seem to have much of an idea. If they did, they didn't talk about it in class.

My brother Lee became a priest, but after a few years he decided that some of the rules, especially celibacy, didn't make sense to him so he resigned, got married, had two daughters and is now enjoying his grand children. After I became involved in the study of the afterlife, he told me that he thought I was more spiritual than he was. I asked him why he said that, and he told me it was because of my search for answers about an afterlife and my attempts to communicate with my son. His spirituality was based more on believing what he had been taught by his religious teachers.

I think one of the reasons I had trouble accepting religious teachings as a young man is because of the seeming randomness of the rules. For those who are not familiar with Catholic dogma, I'll explain. There are venial sins—not real serious—but you ended up in Purgatory instead of Heaven if you died with one of these on your record, and mortal sins—very serious—you not only couldn't get into Heaven but you might go to Hell if you didn't repent properly before you died. For example: when I was growing up, telling a "white lie" was a venial sin, but eating meat on Friday was a mortal sin. We usually had fish on Friday (fish wasn't considered meat). However, the church later made it OK to eat meat on Friday. As a young man I wondered what happened to all those people who ate meat on Friday and died before they changed the rule?

It was the inconsistency of rules like this that caused me to question what I was being taught. In addition, I didn't like hearing that my non-Catholic friends couldn't go to Heaven. An attitude of intolerance combined with a tendency to judge people with other beliefs caused me to reject not only my religion, but organized religion in general. The reason I'm commenting on teachings of the Catholic Church is because that is the religion I am most familiar with. As we continue to learn from world events, the radical fundamentalists of other religions can be equally intolerant to the point of causing the deaths of thousands of innocent people.

I have commented a number of times about my feelings regarding mainstream religion and mainstream science, so I would like to make one thing clear. I have a great deal of respect and admiration for all of the people, both religious and non-religious, who dedicate their lives to helping the needy. These are good people who often sacrifice a great deal in order to help those who are less fortunate. Most people know of Mother Teresa, but there are many thousands of others, from philanthropists to soup kitchen volunteers, who work to improve the well-being of their fellow man. In addition, there are many people whose lives are enhanced by their religious leaders and fellow church members.

I also have a high regard for both doctors and scientists. They are sincere, hard-working people, starting with years of study in school and in most cases continuing throughout their careers. The number of people whose lives are saved or simply improved on a daily basis by our medical personnel is phenomenal. Those involved in the natural sciences have made tremendous contributions to everything from developing cures for a huge variety of diseases to explaining many of the physical aspects of the universe. Those who have chosen the social sciences as a career continue to improve our understanding of the human aspects of our lives.

My argument is with those in mainstream science and religion who insist on promoting outdated and unfounded dogmas. In my opinion, it is time for leaders in the religious community to begin examining what life, and the afterlife, is really about. This may sound a little trite, but one of the main lessons I've had reinforced in my mind since losing my son is that what really matters most in life is caring for other people. It doesn't matter how much "stuff" you accumulate, how important you become, or how reverent you are—what seems to matter most is what kind of person you are.

Here is some good news! A change is taking place within our culture. People are beginning to lose their fear of speaking out about very personal, paranormal events that are having positive life-changing effects on them.

In May, 2012 a book entitled *To Heaven and Back* by Mary C. Neal, M.D. was published. Dr. Neal, an orthopedic surgeon, nearly drowned in a kayaking accident. She was held underwater for an unusually long period of time and before being resuscitated had a near-death experience. As stated on Amazon, "Mary's life has been forever changed by her newfound understanding of her purpose on earth, her awareness of God, her closer relationship with Jesus, and her personal spiritual journey suddenly enhanced by a first-hand experience in heaven." Several years after her near-drowning Dr. Neal's nineteen year old son was killed in an accident.

The cover story in the October 15, 2012 issue of Newsweek tells about a new book entitled *Proof of Heaven: A Neurosurgeon's Journey into the Afterlife* by Eben Alexander, M.D. Dr. Alexander's brain was attacked by a very rare bacterial meningitis and he was in coma for seven days. As stated on Amazon "While his body lay in coma, Alexander journeyed beyond this world and encountered an angelic being who guided him into the deepest realms of super-physical existence. There he met, and spoke with, the Divine source of the universe itself." Dr. Alexander describes his experience from the perspective of a neurosurgeon who, prior to his near-death experience, probably would have attributed what he experienced to a lack of oxygen to his brain. He tells his story, as he says "with the logic and language of the scientist I am."

In October, 2012 I went to a lecture by Dr. Neal given at the South Bay Chapter of the International Association for Near Death Studies (IANDS) in which she described her ordeal and in November, 2012 I attended a conference put on by the Forever Family Foundation at which Dr. Alexander also personally described his experience. What really stands out in my mind is just how profound these experiences were for both doctors. The confidence with which they told their stories made it clear that they had experienced something truly transformative and also something that left them more enthusiastic not only about life, but what follows life—the afterlife. I was also impressed by the dedication they both showed to passing on what they learned during their near-death experiences.

An interesting thing I noticed while reading these two books is that, even though Dr. Neal and Dr. Alexander each experienced common aspects of the NDE, such as being out of their bodies and experiencing "beings of light", in some ways their experiences were quite different. These two people, and many others who have had an NDE, were able to experience the afterlife directly. They emphasize how completely real these experiences are. Although what I have experienced through mediumship is more indirect, to me it is every bit as real.

Both of these books are written by scientists. I would recommend them to anyone interested in two very profound yet believable stories about what awaits us after physical death.

I still struggle to find words that describe what I felt when I lost John. There was no physical damage to my brain—that was still intact. But the part of me that I believe is our essence—our non-physical makeup—received a devastating blow. Soon after John died a doctor who is a good friend told me that he thought I was in clinical depression and that I should get professional help. I did have one session with a psychiatrist, who told me he thought I was doing quite well considering the circumstances. Hearing that helped, but what has helped me more than anything else to recover from this "injury" is finding what I now believe are answers to the question of the meaning and purpose of my life.

I am encouraged that the pendulum is beginning to move back toward the middle. In order for this to continue, it is going to require people on both ends of the spectrum to open their minds. It seems that a majority of scientists, who in my experience perceive themselves as being unbiased, are not able to look at emerging information objectively. I hope that this will change.

On the other hand, many of those involved in the practice of traditional religions remain at the other extreme—back where their predecessors were in the days of Galileo. There are theologians and others within traditional religions, for example those at many of the Catholic Universities, who are working to change the outdated doctrines that are keeping many Catholics from understanding the true meaning of spirituality. One of the hurdles these individuals are faced with is getting the church hierarchy to accept the idea that change is not only necessary but long overdue.

The night John died and I lay in bed wondering where my son had gone, I did not expect to end up with a set of beliefs based on data obtained through the implementation of a new science; one that involves the collection and evaluation of anomalous data. My search began by reading a few books, written by individuals who were not afraid to present concepts well outside of mainstream beliefs. I now know that there were many other books available on the subject, but they were often hidden in the occult section of bookstores. Today there are a large number of books readily available that approach the subject from a variety of viewpoints. In general, they present an uplifting, cheerful view of what happens when we die. I think the fact that these books are becoming more and more popular says that there is a need that is being filled. All that is required is for us to examine these books without all of the biases that have burdened our society for far too long.

When I lost John, one of the things that made it so difficult to bear is that I realized that I would never be able to do the things that he and I discussed over the years—beginning the day we walked up the creek—one of which was to go into some sort of international business together. As I complete this book, I have a very strong feeling that he has been involved with me the whole time and in fact is likely to be the one who has been encouraging me to write it in the first place.

Glossary of Terms

Abnormal. Not normal, average, typical, or usual; deviating from a standard.

Afterlife. The afterlife (also referred to as life after death, the Hereafter, the Next World, or the Other Side) is the belief that a part of, or essence of, or soul of an individual, which carries with it and confers personal identity, survives the death of the body of this world and this lifetime, by natural or supernatural means, in contrast to the belief in eternal oblivion after death.

Akashic Records. The Akashic records (akasha is a Sanskrit word meaning "sky", "space" or "ether") is a term used in theosophy (and Anthroposophy) to describe a compendium of mystical knowledge encoded in a non-physical plane of existence. The Akashic records are described as containing all knowledge of human experience and the history of the cosmos. They are metaphorically described as a library; other analogies commonly found in discourse on the subject include a "universal supercomputer" and the "Mind of God."

Anomalous. Inconsistent with or deviating from what is usual, normal, or expected: irregular, unusual.

Auditory Cortex. The primary auditory cortex is a region of the brain that processes sound and thereby contributes to our ability to hear.

Aura. In parapsychology and many forms of spiritual practice, an aura is a field of subtle, luminous radiation surrounding a person or object (like the halo or aureola in religious art).

Aviary. An aviary is a large enclosure for confining birds. Unlike cages, aviaries allow birds a larger living space where they can fly.

Blind Experiment. A blind or blinded experiment is a scientific experiment where some of the people involved are prevented from knowing certain information that might lead to conscious or subconscious bias on their part, invalidating the results.

Celibacy. Refers to a state of being unmarried, or a state of abstinence from sexual intercourse or the abstention by vow from marriage.

Cerebral Cortex. The cerebral cortex is a sheet of neural tissue that is outermost to the cerebrum of the mammalian brain. This is the gray area of the brain hence the name. This is caused by the nerves that lack insulation. The cerebral cortex covers the cerebrum and cerebellum.

Cerebellum. The cerebellum (Latin for little brain) is a region of the brain that plays an important role in motor control. It may also be involved in some cognitive functions such as attention and language, and in regulating fear and pleasure responses, but its movement-related functions are the most solidly established.

Charades. Charades or charade is a word guessing game. In the form most played today, it is an acting game in which one player acts out a word or phrase, often by miming similar-sounding words, and the other players guess the word or phrase. The idea is to use physical rather than verbal language to convey the meaning to another party

Glossary of Terms

Clairvoyance. The term clairvoyance (from French clair meaning "clear" and voyance meaning "vision") is used to refer to the ability to gain information about an object, person, location or physical event through means other than the known human senses, a form of extra-sensory perception. A person said to have the ability of clairvoyance is referred to as a clairvoyant ("one who sees clearly").

Collective Unconscious. Collective unconscious is a term of analytical psychology, coined by Carl Jung. It is proposed to be a part of the unconscious mind, expressed in humanity and all life forms with nervous systems, and describes how the structure of the psyche autonomously organizes experience. Jung distinguished the collective unconscious from the personal unconscious, in that the personal unconscious is a personal reservoir of experience unique to each individual, while the collective unconscious collects and organizes those personal experiences in a similar way with each member of a particular species.

Consciousness. Consciousness is the quality or state of being aware of an external object or something within oneself. It has been defined as: subjectivity, awareness, the ability to experience or to feel, wakefulness, having a sense of selfhood, and the executive control system of the mind

Dazzle Shot. Defined as "some piece of information—whatever it is TO YOU, that you experience as 'right on' or 'wow' or 'that's my family'." Dr. Gary E. Schwartz, *The Afterlife Experiments*.

Discarnate. Having no physical or material body.

Doge. The elected chief magistrate of the former republics of Venice and Genoa.

Dogmatic. Characterized by an authoritative, arrogant assertion of unproved or unprovable principles. In

Catholicism, a dogmatic definition is an extraordinary infallible statement published by a pope or an ecumenical council concerning a matter of faith or morals, the belief in which the Catholic Church requires of all Catholics (although Christians who are not Catholic do not recognize the Catholic Church's authority in such matters).

Double-Deceased Paradigm. A hypothetical situation in which one deceased person brings another deceased person to a medium and potentially assists in the reading as well.

Drop-in. An uninvited discarnate who 'drops in' during a reading.

Dualism. In philosophy of mind, dualism is the assumption that mental phenomena are, in some respects, non-physical, or that the mind and body are not identical. Thus, it encompasses a set of views about the relationship between mind and matter, and is contrasted with other positions, such as physicalism, in the mind-body problem.

Electromagnetic Field. An electromagnetic field (also EMF or EM field) is a physical field produced by moving electrically charged objects. It affects the behavior of charged objects in the vicinity of the field. The electromagnetic field extends indefinitely throughout space and describes the electromagnetic interaction. It is one of the four fundamental forces of nature (the others are gravitation, the weak interaction, and the strong interaction).

The field can be viewed as the combination of an electric field and a magnetic field. The electric field is produced by stationary charges, and the magnetic field by moving charges (currents); these two are often described as the sources of the field. The way in which charges and currents interact with the electromagnetic field is described by Maxwell's equations and the Lorentz force law.

Glossary of Terms

Electronic Voice Phenomena (EVP). Electronic voice phenomena (EVP) are electronically generated noises that resemble speech, but are supposedly not the result of intentional voice recordings or renderings. Common sources of EVP include static, stray radio transmissions, and background noise. Recordings of EVP are often created from background sound by increasing the gain (i.e. sensitivity) of the recording equipment.

Interest in EVP surrounds claims that it is of paranormal origin, although many occurrences have had natural explanations including apophenia (finding significance in insignificant phenomena), auditory pareidolia (interpreting random sounds as voices in one's own language), equipment artifacts, and hoaxes.

Ethereal. Light, airy, or tenuous; extremely delicate or refined; heavenly or celestial; of or pertaining to the upper regions of space.

Fundamentalism. The demand for a strict adherence to specific theological doctrines usually understood as a reaction against Modernist theology, combined with a vigorous attack on outside threats to their religious culture.

Geocentric. In astronomy, the geocentric model is the superseded theory that the Earth is the center of the universe, and that all other objects orbit around it. This geocentric model served as the predominant cosmological system in many ancient civilizations such as ancient Greece. As such, most Ancient Greek philosophers assumed that the Sun, Moon, stars, and naked eye planets circled the Earth.

Golden Rule. Do unto others as you would have others do unto you.

Heliocentric. Heliocentrism is the astronomical model in which the Earth and planets revolve around a relatively stationary Sun at the center of the Solar System.

Copernican heliocentrism is the name given to the astronomical model developed by Nicolaus Copernicus and published in 1543. It positioned the Sun near the center of the Universe, motionless, with Earth and the other planets rotating around it in circular paths modified by epicycles and at uniform speeds. The Copernican model departed from the Ptolemaic system that prevailed in Western culture for centuries, placing Earth at the center of the Universe, and is often regarded as the launching point to modern astronomy and the Scientific Revolution.

Hell. In many religious traditions, hell is a place of suffering and punishment in the afterlife.

Heresy. Heresy is an opinion held in opposition to that of authority or orthodoxy. It is primarily used in a religious context, but by extension (and with increasing frequency), to secular subjects. The term assumes the existence of an orthodoxy.

Heresy (from Greek αἵρεσις, which originally meant "choice", also referred to that process whereby a young person would examine various philosophies to determine how to live one's life) was redefined by the Catholic church as a belief that conflicted with established Catholic dogma.

Innate. Originating in or arising from the intellect or the constitution of the mind, rather than learned through experience.

Materialism. In philosophy, the theory of materialism holds that the only thing that exists is matter or energy; that all things are composed of material and all phenomena (including consciousness) are the result of material interactions. In other words, matter is the only substance, and reality is identical with the actually occurring states of energy and matter.

Medium (spirituality). A practice in which a person claims to be an intermediary between the physical world and the spiritual world.

Mediumship. Mediumship is defined as the practice of certain people—known as mediums—to mediate communication between spirits of the dead and other human beings. While no evidence has been accepted by the wider scientific community in support of the view that there has been communication between the living and the dead, some parapsychologists say that some of their research suggests that such communication may have taken place.

Metaphor. A metaphor is a literary figure of speech that describes a subject by asserting that it is, on some point of comparison, the same as another otherwise unrelated object. Metaphor is a type of analogy and is closely related to other rhetorical figures of speech that achieve their effects via association, comparison or resemblance including allegory, hyperbole, and simile.

Metaphysics. Metaphysics is a branch of philosophy concerned with explaining the fundamental nature of being and the world.

Mind. The complex of cognitive faculties that enables consciousness, thinking, reasoning, perception, and judgment—a characteristic of human beings, but which also may apply to other life forms. A long tradition of inquiries in philosophy, religion, psychology and cognitive science has sought to develop an understanding of what mind is and what are its distinguishing properties. The main questions regarding the nature of mind is its relation to the physical brain and nervous system—a question which is often framed as the Mind-body problem, which considers whether mind is somehow separate from physical existence (dualism and idealism), deriving from and reducible to physical phenomena such as neurological processes (physicalism), or whether the mind is identical with the brain.

Monism. Monism is a point of view within metaphysics which argues that the variety of existing things in the universe are reducible to one substance or reality and therefore that the fundamental character of the universe is unity. Contrasting with this point of view is dualism which asserts that there are two ultimately irreconcilable substances or realities (with consciousness and/or mind on the one hand and matter on the other) or pluralism which asserts any number of fundamental substances or realities more than two.

Monograph A monograph is a work of writing upon a single subject, usually by a single author. It is often a scholarly essay or learned treatise, and may be released in the manner of a book or journal article. It is by definition a single document that forms a complete text in itself.

Moribund. In terminal decline; lacking vitality or vigor.

Morphic Field. "Morphic field" is a term introduced by Dr. Rupert Sheldrake. He proposes that there is a field within and around a "morphic unit" which organizes its characteristic structure and pattern of activity. According to Sheldrake, the "morphic field" underlies the formation and behavior of "holons" and "morphic units", and can be set up by the repetition of similar acts or thoughts.

Morphic Resonance. Essential to Sheldrake's model is the hypothesis of morphic resonance. This is a feedback mechanism between the field and the corresponding forms of morphic units. The greater the degree of similarity, the greater the resonance, leading to habituation or persistence

Glossary of Terms

of particular forms. So, the existence of a morphic field makes the existence of a new similar form easier.

Motor Cortex. Motor cortex is a term that describes regions of the cerebral cortex involved in the planning, control, and execution of voluntary movements.

Multiplicative. Subject to or of the nature of multiplication.

Natural Sciences. The natural sciences are branches of science that seek to elucidate the rules that govern the natural world by using scientific methods. The term "natural science" is used to distinguish the subject matter from the social sciences. The natural sciences include, but are not limited to: Biology, Chemistry, Physics, Geology, Astronomy, Botany, Mathematics and Engineering.

Near Death Experience. A near-death experience (NDE) refers to a broad range of personal experiences associated with impending death, encompassing multiple possible sensations including detachment from the body; feelings of levitation; extreme fear; total serenity, security, or warmth; the experience of absolute dissolution; and the presence of a light. These phenomena are usually reported after an individual has been pronounced clinically dead or otherwise very close to death, hence the term near-death experience.

Noetic. In traditional philosophy, noëtics (from the Greek νοητικός, noētikos, "mental" from noein "to perceive with the mind" and nous "mind, understanding, intellect") is a branch of metaphysical philosophy concerned with the study of mind and intellect.

Occipital Cortex. That part of the cerebral cortex in either hemisphere of the brain lying in the back of the head.

Occipital Lobe. The occipital lobe is the visual processing center of the mammalian brain containing most of the anatomical region of the visual cortex.

Occult. The word occult comes from the Latin word occultus (clandestine, hidden, secret), referring to "knowledge of the hidden" The word has many uses in the English language, popularly meaning "knowledge of the paranormal", as opposed to "knowledge of the measurable" usually referred to as science.

Out of Body Experience (OBE). An out-of-body experience is an experience that typically involves a sensation of floating outside of one's body and, in some cases, perceiving one's physical body from a place outside one's body (autoscopy).

Paradigm. In science, paradigm describes distinct concepts or thought patterns in any scientific discipline or other epistemological context.

Paranormal. A general term that designates experiences that lie outside the range of normal experience or scientific explanation or that indicates phenomena understood to be outside of science's current ability to explain or measure.

Parochial. A parochial school is a school that provides religious education in addition to conventional education. In a narrower sense, a parochial school is a Christian grammar school or high school which is part of, and run by, a parish.

Past Life Regression. Past life regression is a technique that uses hypnosis to recover what practitioners believe are memories of past lives or incarnations.

Phenomenon. A phenomenon, plural phenomena, is any observable occurrence. Phenomena are often, but not always, understood as 'appearances' or 'experiences'.

Glossary of Terms

Philosophy. Philosophy is the study of general and fundamental problems, such as those connected with reality, existence, knowledge, values, reason, mind, and language. Philosophy is distinguished from other ways of addressing such problems by its critical, generally systematic approach and its reliance on rational argument.

Physicalism. A philosophical position holding that everything which exists is no more extensive than its physical properties; that is, that there are no kinds of things other than physical things.

Platitude. A remark or statement, esp. one with a moral content, that has been used too often to be interesting or thoughtful. The quality of being dull, ordinary, or trite.

Psychic. From the Greek "psychikos—of the mind, mental" is a person who possesses an ability to perceive information hidden from the normal senses through extrasensory perception (ESP), or who is said by others to have such abilities

Purgatory. The condition of purification or temporary punishment by which those who die in a state of grace are believed to be made ready for Heaven.

Quantum. In physics, a quantum (plural: quanta) is the minimum amount of any physical entity involved in an interaction. Behind this, one finds the fundamental notion that a physical property may be "quantized," referred to as "the hypothesis of quantization". This means that the magnitude can take on only certain discrete values A photon is a single quantum of light, and is referred to as a "light quantum." Normally quanta are considered to be discrete packets with energy stored in them.

Reality. In philosophy, reality is the state of things as they actually exist, rather than as they may appear or might be imagined. In a wider definition, reality includes everything that is and has been, whether or not it is observable or comprehensible. A still more broad definition includes everything that has existed, exists, or will exist.

Reincarnation. The religious or philosophical belief that the soul or spirit, after biological death, begins a new life in a new body that may be human, animal or spiritual depending on the moral quality of the previous life's actions. This doctrine is a central tenet of the Indian religions and is a belief that was held by such historic figures as Pythagoras, Plato and Socrates. It is also a common belief of pagan religions such as Druidism, Spiritism, Theosophy, and Eckankar and is found in many tribal societies around the world, in places such as Siberia, West Africa, North America, and Australia

Reverence. Feeling or showing deep and solemn respect; deferential. A feeling of profound awe and respect and often love; veneration.

Sanitarium. An establishment for the medical treatment of people who are convalescing or have a chronic illness.

Scientific Method. Refers to a body of techniques for investigating phenomena, acquiring new knowledge, or correcting and integrating previous knowledge. To be termed scientific, a method of inquiry must be based on empirical and measurable evidence subject to specific principles of reasoning. The Oxford English Dictionary says that scientific method is: "a method or procedure that has characterized natural science since the 17th century, consisting in systematic observation, measurement, and experiment, and the formulation, testing, and modification of hypotheses."

Secondary Visual Cortex. The visual cortex of the brain is the part of the cerebral cortex responsible for processing visual information. It is located in the occipital lobe in the back of the brain. During dreaming, the primary visual cortex is inactive, while the secondary visual cortex remains active.

Shaman. A member of certain tribal societies who acts as a medium between the visible world and an invisible spirit world and who practices magic or sorcery for purposes of healing, divination, and control over natural events.

Social Sciences. A group of academic disciplines that study human aspects of the world. They diverge from the arts and humanities in that the social sciences tend to emphasize the use of the scientific method in the study of humanity, including quantitative and qualitative methods. The social sciences include, but are not limited to: History, Sociology, Economics, Political Science, Psychology, Geography, Education and Law.

Soot. A general term that refers to impure carbon particles resulting from the incomplete combustion of a hydrocarbon.

Spirituality. Refers to an ultimate or an alleged immaterial reality; an inner path enabling a person to discover the essence of his/her being; or the "deepest values and meanings by which people live."

Telepathy. The transmission of information from one person to another without using any of our known sensory channels or physical interaction.

Temporal Lobe. The temporal lobe is involved in auditory perception and is home to the primary auditory cortex. It is also important for the processing of semantics in both speech and vision. The temporal lobe contains the hippocampus and plays a key role in the formation of long-term memory.

Trite. Lacking in freshness or effectiveness because of constant use or excessive repetition.

Truncus Arteriosus. Truncus arteriosus is a rare type of heart disease that occurs at birth (congenital heart disease), in which a single blood vessel (truncus arteriosus) comes out of the right and left ventricles, instead of the normal two vessels (pulmonary artery and aorta).

Tuberculosis (TB). Tuberculosis, MTB, or TB (short for tubercle bacillus) is a common, and in many cases lethal, infectious disease caused by various strains of mycobacteria, usually Mycobacterium tuberculosis. Tuberculosis typically attacks the lungs but can also affect other parts of the body. It is spread through the air when people who have an active TB infection cough, sneeze, or otherwise transmit their saliva through the air.

Veridical. True. Pertaining to an experience, perception, or interpretation that accurately represents reality; as opposed to imaginative, unsubstantiated, illusory, or delusory.

White Lie. A minor or benign falsehood.

Xenoglossy. The putative paranormal phenomenon in which a person is able to speak or write a language he or she could not have acquired by natural means.

APPENDIX A

Attribute Definitions and Scoring Values

**A—Accuracy
(-3 to +3)**

Closeness of a statement to truth.

+3 The statement is clearly correct
+2 The statement is probably correct
+1 The statement is possibly correct
0   I do not know if the statement is correct
-1 The statement is possibly incorrect
-2 The statement is probably incorrect
-3 The statement is clearly incorrect

S—Specificity   Estimated rarity of occurrence of
(0 to +3)   the item in the general population, i.e. it does not apply to many people.

                     3   The information is highly specific
                     2   The information is moderately specific
                     1   The information is only mildly specific
                     0   The information is not at all specific

M—Meaningfulness   The statement is relevant and
(0 to +3)   therefore has value to me, i.e. it does not apply to many situations.

                     3   The statement is highly meaningful to me
                     2   The statement is moderately meaningful
                     1   The statement is slightly meaningful
                     0   The statement is not meaningful to me

APPENDIX B

Book Review
Bill Kaspari

SPIRIT MESSENGER: the remarkable story of a seventh son of a seventh son, by Gordon Smith. Hay House, Inc, Carlsbad, California, 2004, 188 pp, $13.95.

When he was born, the midwife who attended Gordon Smith's birth stated that because he was the seventh child of a seventh child, he was going to be very gifted. It apparently was a prophecy that came true.

Gordon begins the book discussing what seems to be the norm in Western society, describing how he became aware of his psychic abilities around the time he was seven or eight years old but was discouraged by his parents from acknowledging this ability. It is interesting to note that his ability showed up soon after recovering from rheumatic fever.

However, when he was twenty five, the apparition of a close friend appeared to him early one morning as he was awakening. Later that day he found out that his friend had died in a fire that morning. Following the funeral service, his friend's sister asked Gordon to take her to a spiritualist church to see a medium. As a result of this visit, he began to attend medium development meetings at one of the local spiritualist churches in his hometown of Glasgow, Scotland.

In Scotland, people who exhibit some mediumistic ability are encouraged to develop this ability over a period of years, usually by attending group meetings conducted by experienced mediums at their local spiritualist church. After several years of training, Gordon began his career.

In the book, he takes you through his own process of development, beginning with his early appearances at local churches and his gradual progression into advanced mediumship. Albert Best, one of the top mediums in the world at the time, became his mentor and friend. Throughout the book he tells many stories that show his unique ability to not only give meaningful information, but to often give precise names and dates. There are also many anecdotal stories about some of the readings he has given, some humorous and others bizarre.

Gordon's writing style connotes sincerity. He uses simple language, and displays the same qualities he admired in Albert Best—humility and a lack of ego. He does not exhibit the pompous, know-it-all attitude that some TV mediums exhibit, and does not make predictions about future events in people's lives. To this day, he still makes a living as a barber, and says that he does not take money for giving readings, even though he is becoming an internationally known medium. Of course, it remains to be seen if he will continue this way.

One of the main points to be gained from reading this book is that if you have lost a loved one, there is substantial evidence given that supports the concept that not only did they survive the death of their body, but you will see them again. Of course, this is a common theme in most books written by mediums, but if you accept that there is no

Appendix B : Book Review

fraud involved, then it is obvious that Gordon Smith has an exceptional talent.

Because he is becoming a renowned medium, Gordon has been asked to participate, along with other mediums, in a number of research studies conducted under controlled conditions by university research scientists. Preliminary results of these studies demonstrate the astounding accuracy of many of the messages he continually receives from the other side.

In addition to many stories about his mediumship, he presents his views on spirituality and discusses other aspects of paranormal phenomena such as séances, trance mediumship, spirit guides and the survival of animal spirits.

In summary, I found the book to be light-hearted, easy to read and a way to gain insights into the likelihood of whether or not we survive the death of our bodies. I recommend it to anyone who is going through the agony that accompanies the loss of a loved one.

Recommended Reading

Listed below are some of the many books I read as I searched for answers to the question of the survival of consciousness. I have also included a few newer books that may be of interest to my readers. The comments below most book titles were taken from the Amazon website.

Life after Life—Raymond Moody, Jr.

In Life After Life Raymond Moody investigates more than one hundred case studies of people who experienced "clinical death" and were subsequently revived. First published in 1975, this classic exploration of life after death started a revolution in popular attitudes about the afterlife and established Dr. Moody as the world's leading authority in the field of near-death experiences. Life after Life forever changed the way we understand both death—and life—selling millions of copies to a world hungry for a greater understanding of this mysterious phenomenon.

The Afterlife Experiments—Gary E. Schwartz, Ph.D.

Risking his academic reputation, Dr. Gary E. Schwartz asked well-known mediums to become part of a series of experiments to prove, or disprove, the existence of an afterlife. This riveting narrative, with electrifying transcripts, documents stringently monitored experiments in which mediums attempted to contact dead friends and relatives of "sitters" who were masked from view and never spoke, depriving the mediums of any cues.

The Presence of the Past—Rupert Sheldrake

Rupert Sheldrake's theory of morphic resonance challenges the fundamental assumptions of modern science. An accomplished biologist, Sheldrake proposes that all natural systems, from crystals to human society, inherit a collective memory that influences their form and behavior. Rather than being ruled by fixed laws, nature is essentially habitual. The Presence of the Past lays out the evidence for Sheldrake's controversial theory, exploring its implications in the fields of biology, physics, psychology, and sociology. At the same time, Sheldrake delivers a stinging critique of conventional scientific thinking. In place of the mechanistic, neo-Darwinian worldview he offers a new understanding of life, matter, and mind.

Closer to the Light—Melvin Morse M. D.

The skeptics have had their say; now listen to the experts. In hundreds of interviews with children who had once been declared clinically dead, Dr. Morse found that children too young to have absorbed our adult views and ideas of death, share first-hand accounts of out-of-body travel, telepathic communication and encounters with dead friends and relatives. Finally illuminating what it is like to die, here is proof that there is that elusive "something" that survives "bodily death."

Recommended Reading

Spirits—They Are Present—Janet Mayer

Is spontaneously speaking an indigenous Brazilian Rainforest language proof of reincarnation? Or proof of channeling? Decide for yourself as you walk the path of psychic medium Janet Mayer. You'll encounter her life experiences of fear, her path to spiritual awakening, mediumship transformation and her two bouts with cancer. She reveals fascinating stories of spiritual life lessons, clients' stories and signs from the other side showing that death is a transition, not an end.

The Truth About Medium—Gary E. Schwartz, Ph.D.

Every Monday night millions of Americans tune into *Medium*, NBC's new hit drama featuring Allison DuBois, an ordinary woman who helps police solve baffling crimes through her ability to communicate with the dead. What most don't know is that this fictional character is based on a true-life medium named Allison DuBois, who is a consultant to the show. For the past four years, DuBois has been the subject of rigorous scientific experiments conducted at the University of Arizona by Harvard-trained psychologist Gary Schwartz. *The Truth about Medium* chronicles many of those experiments as well as the real-life cases Allison has worked on and reveals hard laboratory evidence that psychic ability and mediumship are real.

The Sacred Promise: How Science Is Discovering Spirit's Collaboration with Us in Our Daily Lives—Gary E. Schwartz, Ph.D.

The Sacred Promise brings us into the laboratory of scientist Dr. Gary Schwartz, where he establishes the existence of Spirit by its own Willful Intent—a proof of concept for deceased spirits. The author takes readers on a personal journey into the world of angels and spirits and reveals their existence and desire to help.

Dr. Schwartz candidly discusses the challenges as well as the rewards of connecting with Spirit. He poses several important questions. What if our feelings of emptiness, loneliness, hopelessness, and meaninglessness are actually fostered by our belief in a "spiritless" Universe? What if our physical hunger is symptomatic of a greater spiritual hunger? What if Spirit is actually all around us, ready to fill us with energy, hope, and direction, if we are ready to ingest it? What if Spirit is like air and water, readily available for us to draw within; that is, if we choose to seek it?

Sacred Promise shows how we can attune ourselves and receive this guidance from Spirit, which is all scientifically documented by Dr. Schwartz' experiments and research. Prepare to suspend your beliefs about spirit.

Matter to Mind to Consciousness: Anatomy of the E.L.F.—T. Lee Baumann, M.D.

In Matter to Mind to Consciousness, T. Lee Baumann M.D. transforms how you have always viewed conscious thought. Through his continued pursuit of science—now to the level of the human brain—Baumann demonstrates both our mind's contributions and vulnerability to the electromagnetic medium surrounding us. Most nerve cells in the outermost layers of the brain end blindly, without any continuing nerve connections. Baumann investigates the possibilities underlying this peculiar medical observation and suggests that these neurons are the very beacons of our human awareness and consciousness. Join the author on a journey into the mystical realm of electromagnetism and the proven phenomena which allow our infinitesimally weak brainwaves to radiate into space and circle the globe several times over. For the first time, an explanation exists, not only for paranormal phenomena, but also for our own human awareness and thought. Again, Dr. Baumann proves that reality IS stranger than fiction.

The Oxford University Press Handbook of Psychology and Spirituality Edited by Lisa J. Miller

Postmaterial spiritual psychology posits that consciousness can contribute to the unfolding of material events and that the human brain can detect broad, non-material communications. In this regard, this emerging field of postmaterial psychology marks a stark departure from psychology's traditional quantum measurements and tenets.

The Oxford Handbook of Psychology and Spirituality codifies the leading empirical evidence in the support and application of postmaterial psychological science. Sections in this volume include:

—personality and social psychology factors and implications
—spiritual development and culture
—spiritual dialogue, prayer, and intention in Western mental health
—Eastern traditions and psychology
—physical health and spirituality
—positive psychology
—scientific advances and applications related to spiritual psychology

With chapters from leading scholars in psychology, medicine, physics, and biology, The Oxford Handbook of Psychology and Spirituality is an interdisciplinary reference for a rapidly emerging approach to contemporary science. This overarching work provides both a foundation and a roadmap for what is truly a new ideological age.

Visions, Trips and Crowded Rooms—David Kessler

David Kessler, one of the most renowned experts on death and grief, takes on three uniquely shared experiences that challenge our ability to explain and fully understand the mystery of our final days. The first is "visions." As the dying lose sight of this world, some people appear to be looking into the world to come.

The second shared experience is getting ready for a "trip." The phenomenon of preparing oneself for a journey isn't new or unusual. In fact, during our loved ones' last hours, they may often think of their impending death as a transition or journey. These trips may seem to us to be all about leaving, but for the dying, they may be more about arriving.

Finally, the third phenomenon is "crowded rooms." The dying often talk about seeing a room full of people, as they constantly repeat the word crowded. In truth, we never die alone. Just as loving hands greeted us when we were born, so will loving arms embrace us when we die.

In the tapestry of life and death, we may begin to see connections to the past that we missed in life. While death may look like a loss to the living, the last hours of a dying person may be filled with fullness rather than emptiness. In this fascinating book, which includes a new Afterword, Kessler brings us stunning stories from the bedsides of the dying that will educate, enlighten, and comfort us all.

The Fun of Dying—Roberta Grimes

Most Americans believe that an afterlife exists, but unfortunately mainstream religions teach a generic faith-based view of it and mainstream science ignores the topic altogether. So people are hungry for concrete facts about the afterlife, and afterlife-related books are perennial best-sellers. Just this year, HarperOne published Evidence of the Afterlife: The Science of Near-Death Experiences by Jeffrey Long and Paul Perry, and Harper Collins published Heaven: Our Enduring Fascination With the Afterlife by Lisa Miller, both to much acclaim. But neither book answers the basic questions: Where is heaven? What is it like? and How does it feel to die? Nor do any of the many books about the afterlife answer three more basic questions: How is a solid afterlife possible? Who will make it there and who won't? and What are the rules that we must follow to have the best chance of getting there? The Fun of Dying offers detailed answers to all these questions and more in an easily read hundred pages. It also presents two study guides one brief, and one much more detailed so readers can follow the author in learning these eternal truths. It turns out that for most of us, death is the best time of our lives!

If you wonder whether death ends life, how it feels to die, or what heaven might be like, this book is for you. If you worry about a lost loved one or fret about the death of a pet, all the answers to your questions are here. And if you are afraid of death, if you worry that your life has no meaning, or if you have given up on religions, then let this book ease some of your fears while it brings new meaning to your life.

Nothing in The Fun of Dying is based on the teachings of any religion. Instead, it draws on more than a century's worth of evidence to explain how dying feels, how it happens, and—most importantly—what comes next. Accounts of near-death experiences are just a small part of the afterlife evidence! A lot of the best death-related evidence was produced in the first half of the 20th century, and it has been ignored ever since by mainstream science and mainstream religions. When it is put together with more recent discoveries, it tells a consistent and amazing story. The Fun of Dying is a complete account of how dying feels and what comes next. Read it, learn the truth, and apply its lessons so you can enjoy your best life forevermore.

The Afterlife Unveiled: What the Dead are Telling Us About Their World—Stafford Betty

What happens to us when we die? Many think of heaven as an unimaginable state of bliss. As for hell, it's far out of proportion to any sin we might have committed and makes a travesty of God. But what if the afterlife was something very different? Three decades of research have taught the author, a world expert in the field of death and afterlife studies, where the most reliable sources are to be found. These accounts are far better developed and more plausible than anything found in the world's scriptures or theologies. We hunger for a reliable revelation telling us that life here and now is meaningful and good, that each of us has an important part to play in its proper unfolding, that we are accountable for all we do, and that the spirit-denying materialism all around us is a mistake. The world ahead, unlike ours, is fascinating and fair. Authentic channels through which the "dead" speak are the closest thing to the voice of God that our planet has.

To Heaven and Back: A Doctor's Extraordinary Account of her Death, Heaven, Angels, and Life Again.: A True Story—Mary C. Neal, M.D.

A kayak accident during a South American adventure takes one woman to heaven—where she experienced God's peace, joy, and angels—and back to life again.

In 1999 in the Los Rios region of southern Chile, orthopedic surgeon, devoted wife, and loving mother Dr. Mary Neal drowned in a kayak accident. While cascading down a waterfall, her kayak became pinned at the bottom and she was immediately and completely submerged. Despite the rescue efforts of her companions, Mary was underwater for too long, and as a result, died.

To Heaven and Back is Mary's remarkable story of her life's spiritual journey and what happened as she moved from life to death to eternal life, and back again. Detailing her feelings and surroundings in heaven, her communication with angels, and her deep sense of sadness when she realized it wasn't her time, Mary shares the captivating experience of her modern-day miracle.

Mary's life has been forever changed by her newfound understanding of her purpose on earth, her awareness of God, her closer relationship with Jesus, and her personal spiritual journey suddenly enhanced by a first-hand experience in heaven. To Heaven and Back will reacquaint you with the hope, wonder, and promise of heaven, while enriching you own faith and walk with God.

Proof of Heaven: A Neurosurgeon's Journey into the Afterlife—Eben Alexander, M.D.

Thousands of people have had near-death experiences, but scientists have argued that they are impossible. Dr. Eben Alexander was one of those scientists. A highly trained neurosurgeon, Alexander knew that NDEs feel real, but are simply fantasies produced by brains under extreme stress.

Then, Dr. Alexander's own brain was attacked by a rare illness. The part of the brain that controls thought and emotion—and in essence makes us human—shut down completely. For seven days he lay in a coma. Then, as his doctors considered stopping treatment, Alexander's eyes popped open. He had come back.

Alexander's recovery is a medical miracle. But the real miracle of his story lies elsewhere. While his body lay in coma, Alexander journeyed beyond this world and encountered an angelic being who guided him into the deepest realms of super-physical existence. There he met, and spoke with, the Divine source of the universe itself.

Among Mediums: A Scientist's Quest for Answers—Julie Beischel, PhD

Can psychic mediums really talk to the dead? Following the suicide of her mother and an evidential mediumship reading, Dr. Julie Beischel forfeited a potentially lucrative career in the pharmaceutical industry to pursue rigorous scientific research with mediums full-time. Among Mediums is an accessible, bite-sized review of her 10-year journey and the answers she discovered along the way. Her writing is concise, non-technical, conversational, and entertaining. In Among Mediums, Dr. Beischel discusses her research investigating the accuracy and specificity of information reported by mediums; unique aspects of mediums' experiences, psychology, and physiology; and the potentially useful social applications of mediumship readings in fields including bereavement, end-of-life care, and forensics. All sales support independent research.

Science Set Free—Rupert Sheldrake

In Science Set Free, Dr. Rupert Sheldrake, one of the world's most innovative scientists, shows the ways in which science is being constricted by assumptions that have, over the years, hardened into dogmas. Such dogmas are not only limiting, but dangerous for the future of humanity.

According to these principles, all of reality is material or physical; the world is a machine, made up of inanimate matter; nature is purposeless; consciousness is nothing but the physical activity of the brain; free will is an illusion; God exists only as an idea in human minds, imprisoned within our skulls.

But should science be a belief-system, or a method of enquiry?

Sheldrake shows that the materialist ideology is moribund; under it's sway, increasingly expensive research is reaping diminishing returns while societies around the world are paying the price.

The Synchronized Universe: New Science of the Paranormal—Dr. Claude Swanson

Dr. Claude Swanson was educated as a physicist at M.I.T. and Princeton University. He wanted to understand the Universe at the deepest level. Then one day he discovered that his scientific education had left out a few things—

The parts left out make up the new scientific revolution. Modern physics has suppressed and ignored the paranormal, but in the laboratory paranormal phenomena are now a proven fact and are beginning to shake the very foundation of physics. The new scientific revolution is changing our understanding of the universe and of ourselves—

Handbook To the Afterlife—Pamela Rae Heath and Jon Klimo

In Handbook to the Afterlife, two seasoned experts with decades of experience working with channeled material describe the stages that spirits go through, focusing on the details that these accounts have in common. Just as life itself has different stages of growth and development, so most accounts of the afterlife are consistent with the authors' view that dying and rebirth are also continuous processes. Beginning with the moment of death itself, progressing through different transitional stages, and ending with the return of spirits to the physical plane, authors Pamela Heath and Jon Klimo define the purposes and pitfalls of each stage. They look at the kinds of adjustment problems that occur in each phase, and how spirits can be helped to move forward. Questions of pain and emotional state at the time of death, karma, and reincarnation are sensitively addressed. The book includes practical techniques for opening up communication with those who have passed on to the other side. While of interest to anyone seeking a general overview of the subject, Handbook to the Afterlife is particularly useful for those dealing with spirits who have not moved on, such as ghosts.

Journey's Out of the Body—Robert A. Monroe

With more than 300,000 copies sold to date, this is the definitive work on the extraordinary phenomenon of out-of-body experiences, by the founder of the internationally known Monroe Institute.

Children Who Remember Previous Lives: A Question of Reincarnation—Ian Stevenson M.D.

This is the revised edition of Dr. Stevenson's 1987 book, summarizing for general readers almost forty years of experience in the study of children who claim to remember previous lives. For many Westerners the idea of reincarnation seems remote and bizarre; it is the author's intent to correct some common misconceptions.

New material relating to birthmarks and birth defects, independent replication studies with a critique of criticisms, and recent developments in genetic study are included. The work gives an overview of the history of the belief in and evidence for reincarnation. Representative cases of children, research methods used, analyses of the cases and of variations due to different cultures, and the explanatory value of the idea of reincarnation for some unsolved problems in psychology and medicine are reviewed.

Soul Shift—Mark Ireland with Foreword by Tricia Robertson

Businessman Mark Ireland's father was Richard Ireland, a deeply spiritual minister and renowned psychic and medium who counted Mae West among his famous clients. While he loved his father, Mark followed a more conventional path in pursuit of mainstream success—until the wrenching death of his youngest son. This unexpected tragedy plunges Mark into the spiritual world of psychics and mediums in a frantic attempt to communicate with the dead. His defenses and pragmatic mindset begin to fade as he remembers premonitions on the day of his son's death. He consults a number of well-known mediums and is struck by the remarkably accurate information their readings provide. Mark first meets with Allison Dubois, the subject of NBC's hit show Medium, and later participates in a single-blind lab experiment with medium Laurie Campbell, filmed for a Discovery Channel feature. He then enters a new dimension of personal paranormal experience, as his own psychic awareness begins to unfold. This dramatic story of a father's unbearable loss and his discovery of life after death offers hope to the bereaved and compelling evidence that death may not be the end.

Your Psychic Potential—Richard Ireland

Known as the "Psychic to the Stars," Richard Ireland counseled celebrities including Mae West, Amanda Blake, and Glenn Ford. Twelve years after Ireland's death in 1992, his son Mark was sent this manuscript, written in 1973. Recently, as Mark Ireland recounts in the foreword, two psychic-mediums with no prior knowledge of the project have received messages suggesting that his father deliberately delayed the book's release until now, when it would reach an audience more receptive to developing their psychic talents.

Your Psychic Potential includes a description of the four spheres/levels of psychic activity, an exploration of the relationship between artistic talents and the psychic, tests and experiments to help unleash psychic ability, a psychic's diet and meditative exercises that support the freer flow of abilities, and tools to counter inhibitory fears. Anyone interested in discovering their extrasensory talents and achieving conscious control over them is sure to find this an indispensable guide.

Spirit Messenger—Gordon Smith

Gordon Smith, a native of Glasgow, Scotland, is the proverbial seventh son of a seventh son. His outstanding abilities as a medium or messenger from the spirit world have led him to be dubbed "The UK's most accurate medium". Gordon has brought comfort and healing to thousands of people through the messages he is delivered from the other side. He has traveled to many parts of the world to appear before audiences, has read for many celebrities, and has been featured in television documentaries, but his feet have remained firmly on the ground. Refusing to ever charge for a private reading, Gordon still works in his barbershop in the west end of Glasgow. Gordon has proven his abilities to scientists exploring the nature of mediumship, and has astounded them with the consistency and accuracy of his messages. In this often humorous book, Gordon tells his own story of life as a messenger for the spirit world and shares his commonsense advice for avoiding charlatans and finding a true connection to Spirit.

Recommended Reading

Where God Lives—Melvin Morse M. D.

Is there proof that "near death" and other spiritual experiences can cure afflictions of the body, mind, and spirit? Are there simple ways to tap into a "universal power source" that spiritual masters call enlightenment? Is there scientific evidence of life after death that is being overlooked by skeptics? Is there scientific proof of a spot in our brains that communicates with God and the universe? Pediatrician Melvin Morse believes the answer to all these questions is yes. Shedding new light on the links between science and mysticism, *Where God Lives* not only reveals the area of the brain that is our biological link to the universe, but also shows us the secret of tapping into the universal energy to achieve healing, personal peace, and transcendence. Filled with moving case histories, *Where God Lives* applies the rigor of science to the study of the spiritual to prove once and for all the existence of life after death.

Reunions—Raymond Moody, Jr.

A collection of the experiences of men and women who have communicated with the dead using the easy-to-learn techniques developed by Dr. Raymond Moody. As proof of life after death, these stunning testimonials promise to launch even more research and give comfort to people around the world.

Many Lives, Many Masters—Brian L. Weiss, M.D.

As a traditional psychotherapist, Dr. Brian Weiss was astonished and skeptical when one of his patients began recalling past-life traumas that seemed to hold the key to her recurring nightmares and anxiety attacks. His skepticism was eroded, however, when she began to channel messages from the "space between lives," which contained remarkable revelations about Dr. Weiss's family and his dead son. Using past-life therapy, he was able to cure the patient and embark on a new, more meaningful phase of his own career.

Messages—George Dalzell

If you had asked George Dalzell, a professional psychiatric social worker, a few years ago if he thought he'd be talking with the dead, he would have said, 'no way.' Yet, when his close friend Michael, an airline purser from Germany, was killed, things started happening that permanently shifted Dalzell's "perspective on reality."

As a therapist, Dalzell had counseled people who claimed to hear voices. Now, Dalzell was hearing a voice himself, and it was that of Michael. The voice revealed information about Michael's private life and possessions. Other phenomena followed, including apparitions of Michael, house lights that blinked on and off, alarm clocks that got moved, and rose petals left in the pattern of an angel. "Michael appeared to have brought evidence that there was a dimension beyond our five senses, and that he could keep the channels open", writes the author.

The Ghost Detectives—Loyd Auerbach & Annette Martin

Revealing a side of the famed city that tourists rarely experience, this handbook uncovers a hidden realm of ghosts, apparitions, and paranormal phenomena in San Francisco. The guide delves into the haunted hotspots that unsuspectedly lie in the city's most famous landmarks and neighborhoods, including Alcatraz, Chinatown, and the Presidio, while directions to each hair-raising location are provided, encouraging adventurous sightseers to seek out their own ghostly encounters. With the history of each frightening locale, the probable life stories of their resident spirits, and actual transcripts of their conversations with a psychic, this supernatural study delivers a realistic feel for encountering the uncanny.

The Airmen Who Would Not Die—John G. Fuller

In 1928, as the *Graf Zeppelin* prepared to fly around the world, the British raced to complete the luxurious R 101 airship that was to revolutionize air travel. In spite of severe structural problems, the government had decided that the take-off date could not be postponed—for British pride was at stake . . .

Nor would they heed the detailed, fearsome warning from a dead World War One ace—a warning from the life beyond . . .

The R 101 plunged to the ground on the French side of the Channel . . . and two days later, during a séance, the commander of the ill-fated airship related in ghastly detail, R 101's tragic end . . .

Bibliography

Moody, Raymond. *Life After Life.* New York: Bantam Books, 1975.

Morse, Melvin. *Closer to the Light.* New York: Random House, 1990.

Monroe, Robert A. *Journeys Out of the Body*: Doubleday, 1971

Schwartz, Gary E. *The Afterlife Experiments.* New York: Simon & Schuster, 2002.

Sheldrake, Rupert. *The Presence of the Past.* New York: Random House, 1989.

Schwartz, Gary E. The Truth about Medium. Virginia: Hampton Roads, 2005.

Schwartz, Gary E. The Sacred Promise. Simon & Schuster, 2011

Baumann, T. Lee. *Matter To Mind To Consciousness: Anatomy of the E.L.F.,* North Charleston, SC, 2011

Kessler, David. *Visions, Trips, and Crowded Rooms.* California: Hay House, 2010.

Grimes, Roberta. *The Fun of Dying.* Greater Reality Publications, 2010.

Betty, Stafford. *The Afterlife Unveiled.* O-Books, Washington, D.C., 2011.

Neal, Mary C. *To Heaven and Back.* WaterBrook Press, Colorado Springs, CO, 2012

Alexander, Eben. *Proof of Heaven.* Simon and Schuster, New York, NY, 2012

Heath, Pamela Rae & Klimo, Jon. *Handbook to the Afterlife.* North Atlantic Books, California, 2010.

Sheldrake, Rupert. *Science Set Free.* Deepak Chopra Books, Random House, New York, NY, 2012

Swanson, Claude. *The Synchronized Universe.* Poseidia Press, Tucson, AZ, 2003

Morse, Melvin. *Where God Lives.* HarperOne, 2000.

Website Addresses & Information

1. Pat McAnaney: http://www.psychicpat.com/bio.html

2. University of Arizona Veritas Research Program

   Julie Beischel, PhD and Gary E. Schwartz, PhD

   Rhine Research Conference: "Consciousness Today" March 23-25, 2007

   Methodological Advances in Laboratory-Based Mediumship Research http://windbridge.academia.edu/JulieBeischel/Papers/566943/Methodological_advances_in_laboratory—based_mediumship_research

3. University of Arizona Veritas Research Program: Closing of the program. http://lach.web.arizona.edu/veritas_research_program

4. Forever Family Foundation: www.foreverfamilyfoundation.org

5. The Windbridge Institute: http://www.windbridge.org/

   Peer-Reviewed Research Papers: http://windbridge.org/publications.htm

Julie Beischel's paper: Beischel, J. (2007). Contemporary methods used in laboratory-based mediumship research. Journal of Parapsychology, 71, 37-68.

Here is a link to the PDF version of the paper: http://windbridge.org/papers/BeischelJP71Methods.pdf

6. Rupert Sheldrake: http://www.sheldrake.org/homepage.html

7. Melvin Morse: Spiritualscientific.com

8. Ian Stevenson: http://www.sinor.ru/~che/birthmarks.htm http://www.afterlife101.com/Xenoglossy.html http://www.ial.goldthread.com/kidspage/stevenson.html

9. David Kessler http://grief.com/david-kessler-bio

10. T. Lee Baumann: https://profiles.google.com/tleebaumann/about

11. Ray Hyman: http://www.csicop.org/si/show/how_not_to_test_mediums_critiquing_the_afterlife_experiments//

12. Michael Shermer: http://www.skeptic.com/reading_room/the-great-afterlife-debate/

13. Roberta Grimes http://afterlifeforums.com/entry.php/161-Roberta-Grimes-Talks-About-The-Fun-of-Dying

14. Stafford Betty http://www.csub.edu/~sbetty

15. Religious Tolerance: http://www.religioustolerance.org/var_rel.htm

16. Telepathy: http://en.wikipedia.org/wiki/Ganzfeld_experiment

Website Addresses & Information

17. Monroe Institute: http://www.monroeinstitute.org

18. Washington Post Article: http://www.washingtonpost.com/business/technology/steve-jobss-last-words-oh-wow-oh-wow-oh wow/2011/10/31/gIQA3vKCZM_story.html

19. Institute of Noetic Sciences http://noetic.org/

20. The Andromeda Galaxy Image used for the cover: http://www.nasa.gov/images/content/386913main_Swift_M31_large_UV.jpg

21. Understanding Afterlife - a general afterlife discussion group. www.UnderstandingAfterlife.com.

Acknowledgements

When I think of the people who are my greatest inspiration, I think of my children and grandchildren. This is not just another platitude. As each of my children came along, they made my life more enjoyable. Watching them grow and learn gave me great pleasure, and was a learning process for me as well. Being able to live this all over again with my grandchildren is wonderful. Kelsey, Olivia and Kamlyn, I love each of you far beyond what human language can express.

Diane—you and I have been through a lot together. You know as well as anyone what we lost when we lost John, but as we have said many times, we would rather have had him in our lives for those twenty-two years than to never have had him at all. Thanks for helping me to keep the memories of John and Ricky alive in your writings, in all of the pictures you have taken and in our daily conversations, and thanks for all of the support you've given me while I was writing this book.

Angie—I can not thank you enough for taking the time to do such a thorough job of critiquing my writing. Your editing was invaluable. You have a special gift for putting thoughts and feelings into words. Your suggestions helped me convert some of the anger and resentment I felt when writing certain sections into a much more positive way of getting my message across.

Becky—without your comments this book would not have been complete. Thanks for reading my manuscript and for encouraging me to make this book much more personable than it was originally. Your suggestions caused me to make some badly needed changes.

Jim—the suggestions you made were outstanding. You pointed out some important items that were missing and others that needed to be deleted. Thanks for that and for your support.

Wayne and Lee—thanks for being such great brothers and friends over the years and for your encouragement and helpful comments while I was writing this book.

To my brother-in-law David Littlejohn—thanks for your advice and many helpful hints related to writing this book.

Gary Schwartz—thanks for taking the time to meet me that day at your hotel in San Francisco. It's hard to imagine where my search might have taken me if we had not worked together furthering the study of afterlife science. You have helped me to achieve the paradigm shift in my thinking that has changed my life. Also, thanks for reviewing my manuscript and for writing the Foreword.

Janet Mayer—you are a very special lady. You have helped open my eyes to the unseen world that I now believe awaits us all. Thanks for being so receptive to my sons, my father and all of the others who showed up in my readings. Thanks for not only providing your readings but for helping me, and hopefully many more people, to accept and understand mediumship.

Allison DuBois, Laurie Campbell, Mary Ochino, Gordon Smith, and George Dalzell, thanks for providing readings for me. I would like to have included more readings than I did, and all of the ones you provided were certainly meaningful.

Acknowledgements

Julie Beischel & Mark Boccuzzi.—Julie, my thanks to you begin with the day you picked me up at the airport in Tucson, and continue to this day. Your professionalism, lucid mind and helpful suggestions made my time at the U of A lab not only informative but enjoyable. I am proud of you and Mark for having the courage to continue the work needed to help advance our understanding of paranormal phenomena.

Bob & Phran Ginsberg—we share a bond that I know we both wish had not been necessary because it was forged due to the loss of our children. You are two of the people in the "most admired" category of my mind. What you have done, and continue to do through the Forever Family Foundation to help those grieving the loss of a loved one, is remarkable.

Maisie and Chris MacDonald—I know what a wonderful, cheerful and bright young lady you lost when Amanda died. I will always treasure memories of the good times we shared with your family, especially the father/daughter dance I was so fortunate to attend with Amanda and Angie. Thanks for being such wonderful friends.

Arthur Avary—you are a truly gifted artist. I don't think I could have found a person who was able to take the ideas I had for cover art and not only create the art but also the feeling I was trying to portray. Thanks also for being John's good friend.

Dr. Stafford Betty, Dr. H. Bruce Greyson and Tricia Robertson—thanks for writing such compelling endorsements for this book. It is an honor to have one from each of you.

Dinah James—thanks for giving me the book *Life After Life*. You helped me to start my search in the only way I was able to accept.

Barb Schneiderman—thanks for so generously setting up the reading with Pat McAnaney.

Pat McAnaney—thanks for helping to open my eyes to that Unseen World. When you first provided a glimpse of it, my eyes and my mind were still tightly closed, but you helped convince me that there is more than what "meets the eye". Thanks also for your support and your positive attitude.

Al Brack—thanks for recommending the book *The Presence of the Past*. That book was instrumental in helping me to understand the limitations in my thinking.

Mark Ireland—thanks for all of the helpful hints you provided, especially your many suggestions for ways to promote my book and the message behind it. Also, thanks for all you are doing to help those who have lost a loved one. You are a good friend.

The author is grateful for permission to use excerpts from the following works:

Baumann, T. Lee. *Matter to Mind to Consciousness: Anatomy of the E.L.F.*, North Charleston, SC, 2011.

Betty, Stafford. *The Afterlife Unveiled*, O-Books, Washington, D.C., 2011.

Grimes, Roberta. *The Fun of Dying.* Greater Reality Publications, 2010.

Kessler, David. *Visions, Trips, and Crowded Rooms.* California: Hay House, 2010.

Sheldrake, Rupert. *The Presence of the Past.* New York: Random House, 1989.

I'd like to thank all of the people on the TestingMediums website who have contributed to my education, including but certainly not limited to: the late Steve Grenard, Peter Hayes,

Acknowledgements

David Haith, Tricia Robertson, Mark Ireland, Don Watson, and many others too numerous to mention.

The people at AuthorHouse—Valerie Raines, April Ross, Jay Roberts, Amanda Miller and all of those who have made the publication of this book possible—thanks for being so available and helpful.

Google. I used your amazing search engine repeatedly.

The following Internet based sites provided invaluable aid:

Wikipedia. All of you who have created and contributed to this remarkable website have done a fantastic service for all of us.

Amazon. For all of the book descriptions.

Brainy Quotes. Thanks for providing a source of quotations.

Religious Tolerance. For the information about religions.

Index

A

Aaron 25
abortion 8
A Brief History of Time 36
accuracy 39, 61, 70, 129, 191, 206
accurate 39, 62, 64, 205, 206
Adam Sedgwick 115
AD Mattson 150
after-death communication 144
after-death levels 144
afterlife phenomena 4, 78, 124, 126, 137
Afterworld 145, 146, 148
Akashic 101, 147, 173
Albert Einstein 15, 99
Aldous Huxley 71
Alexander (Eben Alexander, M. D., Dr. Alexander) 168, 201
Allah 120
Allison DuBois 40, 51, 71, 195, 218
Alzheimer's disease 131
Amanda 25, 47, 205
Amazon 38, 103, 193, 211, 221
amino acids 138

Among Mediums: A Scientist's Quest for Answers 91
anesthetic agents 22
Angie 12, 25, 47, 217
Anglican 149
animals 16, 43, 63, 107
anniversary 76
anomalous data 170
antenna 105, 108, 109, 110, 132
antidepressants 22
aorta 11, 186
apparition 119, 158, 189
Aramaic name for Jesus 145
Argentina 38
Aristotle 25
arrogance 120
artistic or athletic skill 90
A SAMPLE READING 61
Asking Questions experiment 49
astral 146, 148, 149, 150
atheistic 142
atoms 105
auditory or visual cortex 107
aura 27, 174
aviary 97, 174

B

baby shower 77

Bailey 50
BALD EAGLE 68
Baptist 149
Barb 162
Baumann, T. Lee (Dr. Baumann, T. Lee Baumann, M.D.) 108, 110, 152, 196, 197, 211, 214
Becky 12, 25, 33, 77, 86
beginnings of life 139
Beischel, Julie (Dr. Beischel, Julie Beischel, Ph.D., Dr. Julie Beischel) 49, 51, 53, 69, 90, 91, 213, 214, 219
bereaved 90, 92, 205
Betty, Stafford 145, 150, 151, 152, 156, 212
Betty, StateStafford 200, 214
biochemistry 103
birth defects 100
birthmarks 100, 214
Boccuzzi, Mark 54, 90, 219
brain 43, 98, 101, 102, 105, 106, 107, 108, 109, 110, 113, 116, 131, 132, 133, 161, 162, 169, 173, 174, 179, 184, 196, 207
Brazil 38
Buddha 1
Buddhism 142, 153, 154
Buddhist 149, 154
bump-on-the-head 128

C

California 10, 163, 189, 211, 212, 220
California State University in Bakersfield 145
Cambridge University 103
Camden Primary Care Trust 119
Canaries 97

Carl Sagan 103
Cartesian coordinate system 116
casket 159
Casper 76, 82
categories of memories 101
Catholic 2, 149, 163, 165, 166, 176, 178
Catholic Universities 170
cerebral and cerebellar cortices 108
certification process 92
chaplains 119
charades 49, 54, 55, 61, 65, 68, 69, 127, 128, 151
Chicago 8, 9
Children's Hospital 22
Chopra, Deepak (Dr. Chopra) 130, 131, 132, 133
Chris 153
Christ 153, 165
Christianity 145, 153
Christians 117, 176
Christian scriptures 145
Christmas 8, 72, 77, 122
church 2, 8, 159, 166, 178, 189, 190
church hierarchy 170
CIA 96
clairvoyant 27, 175
Claremont Graduate University 130
Closer To The Light 20
coincidence 74, 97
cold reading 126
collective unconscious 101, 175
color auras 27
compassionate 38
consciousness 2, 5, 23, 91, 102, 109, 116, 123, 124, 125, 132, 138, 178, 179, 180, 193, 196

Index

Conservative Judaism 154
continuum 156
Copernicus (Nicolaus Copernicus) 2, 16, 178
cortical surface 109
cosmos 139, 173
counselors 119
C. P 28
Creator 148
creek 34, 35, 97, 171
critic 125
Crowded Rooms 118, 211, 220
Cuesta College 118
culture 1, 2, 13, 16, 17, 26, 115, 140, 177, 178

D

dad 7, 8, 9, 34, 55, 60, 63, 64, 65, 68, 73, 74, 77, 85, 95, 96, 127, 128, 151, 153, 156, 159, 160
Darline 38
Darwin, Charles 139
Daryl Bem 152
data 50, 62, 74, 124, 127, 128, 129, 135, 139, 170
David Haith 79
dazzle shot 128
dazzle shots 128
DBP (deathbed phenomena) are 119
Dean Radin 152
death ix, 2, 3, 4, 5, 13, 17, 18, 19, 20, 21, 22, 25, 26, 28, 29, 37, 38, 41, 48, 76, 77, 78, 99, 102, 103, 116, 117, 118, 119, 120, 122, 123, 124, 131, 139, 140, 142, 143, 148, 151, 156, 158, 159, 160, 161, 173, 181, 184, 190, 191, 193, 194, 195, 198, 199, 200, 203, 205, 207
deceased 20, 39, 48, 54, 55, 59, 68, 70, 75, 77, 78, 113, 128, 158, 159, 195
de-oxygenated blood 11
Descartes 116, 117
destroyer 1, 10
Diane 10, 11, 12, 13, 25, 26, 28, 38, 50, 55, 68, 72, 73, 75, 77, 86, 96, 101, 117, 158, 159, 160
Dilantin 22
discarnate 65, 69, 118, 161, 176
discarnates 49, 50, 54, 61, 69, 75, 121, 122, 142, 161
Divine 146
DNA 101, 131, 132, 138
doctors 20, 22, 106, 119
doctrines 170, 177
dogma 155, 166
dogmatic 2, 3, 176
double and triple blind studies 69
double blind experiment 70
drugs 22, 119, 133
dualism 115, 116, 139, 176, 180

E

earth-status 144
Eastern religions 142
East-West bookstore 75
Edgar Mitchell 94
Edith Fiore, PhD 32
Egyptians 115
Einstein 16
Electrical Engineering 1, 10
electro-magnetic fields 104, 105
Electronic Voice Phenomena (EVP) 163

Ellen 35
emotions 13, 49, 117, 132, 147
energy 5, 27, 58, 59, 77, 82,
    83, 106, 138, 141, 143,
    178, 183, 196, 207
enlightenment 143, 207
epileptics 107
epiphenomenon 116
Eric C 42
ERIC CLAPTON 42
ESP 85, 132, 183
essence 131, 161, 162, 169,
    173, 185
ethereal 156
evaluating a reading 40
Evidence of Mind 132
evolution 132, 138, 139
experiment 10, 49, 84, 121,
    129, 174, 184, 205
extremely low frequency
    (ELF) 109

F

faith 57, 120, 142, 153, 176, 198
farm animals 56, 63
father 8, 42, 47, 55, 56, 58, 63,
    64, 77, 82, 86, 96, 205
FBI 96
fear 4, 13, 23, 106, 117, 174,
    181, 195
feelings ix, 12, 28, 49, 55, 83,
    84, 117, 132, 157, 162,
    181, 196
fishbowl 78
foreign language 57, 63
Forever Family Foundation 76,
    90, 91, 163, 213
fortune tellers 26
frauds 89, 124
*Full Facts Book of Cold
    Reading* 134

fundamentalists 167

G

galaxies 105
Galileo 2, 16, 170
geocentric 2, 177
George Anderson 125
Germany 160, 208
Ginsberg 50, 90
God 2, 110, 142, 145, 153, 154,
    165, 173, 200, 207, 212
gold country 34
golden rule 17
gold locket 75
Google 94, 96, 153, 221
Gospels 142
Grand Canyon 34
gravitational field 104
gravity 141
Greece 32, 115, 177
Grimes, Roberta 139, 140,
    141, 142, 144, 145, 150,
    151, 152, 156, 198, 212,
    214, 220

H

Hactic 58, 60
Haldol 22
hallucinations 107, 119
Handbook To The Afterlife 155
Harvard 38, 130, 195
Hawaii 34, 58, 64, 65, 95,
    127, 159
Heaven 95, 153, 166, 183, 199
heliocentric 2
Hell 166, 178
hellish regions 149
Henri Poincare 157
Henry David Thoreau 79
hereafter 95, 117, 138
Hindu 149, 154

Index

Hinduism 142, 153, 154
hospices 119, 146
hospitals 119
How Not to Test Mediums 125
Humanistic Judaism 155
Hyman, Ray (Dr. Hyman, Hyman) 125, 126, 127, 128, 129, 130, 134, 214
hypnotism 31
hypothesis 4, 21, 129, 135, 180, 183
hypothetical signals 161

I

Ian Rowland 133
identifying code 161
images 33, 40, 47, 48, 55, 60, 65, 66, 84, 105, 108, 121, 161
influence 10, 28, 97, 98, 104, 113
Information Fields 132
information source 107
infra-red 5, 89
insects 107
Institute of Noetic Sciences 94, 134
International Association for Near Death Studies (IANDS) 169
internet 40
Islam 153, 154
Italy 34

J

Janet Mayer 40, 41, 54, 76, 79, 80, 118, 128, 151, 195, 218
JAZZ BAND 42
JAZZ FESTIVALS 42
jewel 59, 60
Jewish 120
Jim 10, 25
Jiu Jitsu 10
John vii, 12, 13, 18, 19, 20, 25, 26, 27, 28, 32, 33, 34, 35, 37, 40, 41, 42, 44, 47, 48, 55, 64, 68, 73, 75, 76, 77, 78, 80, 81, 82, 83, 86, 96, 97, 103, 118, 151, 158, 159, 160, 161, 169, 170, 171
John Edward 125
John Gribbin 36
John Hopkins University 125
John Kaspari Fund 49
Jon Klimo 155, 203, 204
Journeys Out of the Body 29, 162, 211
Judaism 153, 154, 155
Jung 101, 116, 175

K

Katie 20, 21, 22, 133
Kepler 16
Kessler, David 117, 119, 120, 198, 211, 214, 220
KNIGHTS OF COLUMBUS. 60

L

Laurie Campbell 96, 205
Lee 8, 9, 166
level of existence 156
liberal arts 15
Life After Death: The Burden of Proof 131
Life After Life 19, 193, 211, 219
lucid dreams 151, 159, 161
Lucid Dreams 159

M

Ma 162
magician 31, 133

magnetic fields 133
mainstream religion 139, 140, 142
mainstream science 2, 3, 5, 51, 126, 139, 140, 142, 198
Many Lives, Many Masters 101, 207
Margaret Mead 165
mark 134, 149
Martin Luther King, Jr 19
Mary Ochino 40
materialism 115, 116, 138, 178, 200
materialistic 115, 138, 139
materialization 121
material world 5, 145
matter 5, 90, 104, 106, 116, 127, 138, 141, 161, 167, 176, 178, 180, 194
Matter to Mind to Consciousness 108, 196
meaningful 41, 50, 62, 64, 128, 188, 190, 200
mechanistic 104, 117, 194
medium 26, 39, 40, 41, 48, 54, 60, 61, 62, 64, 68, 69, 70, 75, 76, 86, 87, 90, 91, 96, 125, 127, 129, 131, 147, 151, 185, 189, 190, 191, 195, 196, 205, 206
mediumship skills 90
memorial service 12, 159
memories 13, 31, 34, 101, 102, 105, 106, 107, 108, 131, 132, 133, 182
Mercy Hospital 8
Mesa Verde 34
metaphysical 140
Michael Persinger 133
Michael Talbot 36

Miller (Lisa Miller, PhD) 110, 197, 199
mind 4, 11, 16, 22, 29, 32, 49, 52, 55, 57, 61, 72, 77, 97, 104, 106, 116, 117, 130, 131, 132, 151, 157, 158, 159, 175, 176, 179, 180, 183, 194, 196, 207
Minds, Brains and Memories 105
Missionary spirits 149
mom 8, 9, 96, 153, 158, 159
monism 115
Monroe (Robert A Monroe, Robert Monroe) 29, 30, 162, 204, 211, 215
mood elevators 22
Moody, Jr. M.D, Raymond (Dr. Moody) 19
Mormon 149
morphic fields 104
Morphic Resonance 132, 180
Morse, Melvin (Melvin Morse, M.D., Dr. Morse) 20, 21, 110, 133, 194, 207, 211, 212, 214
mortal sin. 166
mother 7, 8, 45, 48, 56, 63, 64, 65, 68, 77, 95
Mother Teresa 167
motor cortex 107
motorcycle 12, 28, 38, 47, 48, 86, 160
motorcycles 38
Mountain View, CA 75
Mount Vernon 9
MUSIC SCHOLARSHIP 43
Muslim 149
Muslims 117
MVP 25

N

narcotics 22
National Institute of Health (NIH) 52
Natural Sciences 167, 181
Neal (Mary C. Neal, M.D., Dr. Neal 168, 169, 200
Near Death Experience (NDE) 19, 20, 29, 162, 181
neo-Darwinian 104
Neonatal Intensive Care Unit (NICU) 11
neuron 109
neuronal record 108
neurons 109, 113, 131, 196
neurophysiologist 107
new age person 35
Newton 16
Newtonian physics 140
Nido Qubein 7
non-physical 3, 4, 5, 121, 132, 169, 173, 176
North American Jews 155
nurses 119
nursing homes 119

O

objective 61, 69, 74, 124, 129, 130, 135, 144
objectivity 10, 29, 62, 157
occult 23, 170, 182
old soul 27, 29
omniscient 147
open mind 4, 5, 32, 41, 52, 74, 123, 138, 145, 157
OReilly 38
organelles 109
origin-of-life theories 138
Orthodox Judaism 155

other side 33, 62, 67, 68, 69, 78, 80, 121, 161, 191, 195, 204, 206
Out of Body Experience, (OBE) 29, 162
oxygenated blood 11

P

Packard Children's Hospital 25
painkillers 22
palliative (end-of-life) care facilities 119
Pamela Rae Heath 155, 203
paradigm ix, 2, 5, 16, 182
Parakeets 97
Paramed (Paramed Technology) 37, 38
paranormal 5, 26, 27, 31, 36, 52, 53, 90, 106, 123, 124, 126, 130, 134, 135, 156, 159, 161, 177, 182, 186, 191, 197, 205
Paranormal Phenomena 103
parapsychology 125, 174
Paris 34
particles of matter 141
Past Life Regression 31
past life regression (past-life regression) 36, 100
Pat McAnaney 26, 27, 28, 87, 213
Paul Davies 36
Paul N. Temple 94
peace 46, 59, 67, 158, 207
peer review 134
pendant 75
pendulum 3, 51, 170
perceptions 36, 40, 89
peripheral nerve terminals 109
personality 55, 62, 65, 128, 131

Peter Hayes 53, 61, 80, 220
Pharaohs 115
pharmacology 22
phenomena 3, 4, 5, 20, 21, 26, 31, 36, 52, 53, 74, 89, 106, 119, 121, 123, 124, 126, 130, 135, 137, 138, 159, 161, 176, 177, 178, 180, 181, 182, 184, 191, 196
philosopher 116
Philosophical skepticism 123, 135
physicalism 115, 176, 180
physical mediumship 121
physical realm 95
physical world ix, 3, 4, 5, 16, 26, 48, 105, 122, 125, 134, 140, 179
physics 103, 140, 141, 183, 194
piano 45, 48
pilot study 76, 119
Pisa 34
placeCityCasper 79, 80, 81
placeCitySan Francisco 39
placeCityStevenson, StateMD, Ian (Dr. Stevenson, Dr. Ian Stevenson) 100
placeCityTucson 49
placeCityWeirton 9
placecountry-regionCuba 1
placecountry-regionFormosa 1
placecountry-regionJapan 1
placeHong Kong 1
placeLake Shasta 34
placePlaceTypeUniversity of PlaceNameColorado 10
placeStateAlaska 73
Pope 2
poverty 57, 64, 65, 96, 127
Prayer and Healing Studies 132
primordial chemical soup 138

Proof of Heaven: A Neurosurgeons Journey into the Afterlife 168
proteins 131, 138
psychic 26, 35, 85, 97, 134, 189, 195, 205
Psychic Mediumship 132
Psychic Reading 26
psychics 36, 54, 89, 205
psychologists 119
psychology 38, 101, 103, 116, 125, 130, 133, 175, 179, 194
pulmonary artery 11, 186
Purgatory 166

Q

quanta 141, 183
Quantum Consciousness 132
quantum physics 140

R

radar 89
rainbow 45
Ralph Waldo Emerson 53
rare 62, 64, 186
rats 97
reading 4, 19, 20, 26, 28, 29, 31, 35, 36, 40, 41, 48, 49, 50, 54, 55, 59, 60, 61, 63, 64, 66, 68, 70, 71, 73, 74, 75, 76, 77, 79, 80, 82, 83, 84, 86, 106, 118, 122, 126, 127, 128, 129, 133, 145, 151, 157, 161, 165, 170, 176, 190, 206, 214
reality ix, 23, 104, 116, 126, 138, 141, 156, 179, 180, 185, 186, 197, 208
rebirth 154
Reconstructionist Judaism 155

Reform Judaism 155
reincarnation 31, 99, 100, 101, 150, 154, 195, 204
religion 3, 17, 18, 142, 153, 154, 163, 166, 179
RELIGIOUS 57, 60
religious bias 36
religious schooling 156
Religious Tolerance 153, 214
Ricky vii, 11, 12, 17, 163, 164
RNA 138
Robert Frost 137
Robert Monroe 204, 211
Robertson/Roy 66
robot 117
Roman Catholic 153

S

sanitarium 7, 8, 159
SCENE OF THE GARDEN 58
Schwartz, Gary (Dr. Schwartz, Gary E. Schwartz, Ph.D., Dr. Gary Schwartz, Dr. Gary E. Schwartz) 38, 39, 40, 41, 47, 49, 51, 52, 53, 54, 55, 69, 70, 72, 76, 95, 125, 126, 127, 129, 130, 175, 193, 195, 196, 211, 213, 218
science-based critiques 125
Science Set Free 135, 202
scientific method 10, 17, 152, 184
Scientists 3
Scole Group 121
scoring 40, 49, 50, 54, 60, 61, 63, 65, 66, 69, 70, 126, 128
Scoring Criteria 69
Scotland 75, 190, 206
scriptures 2, 17, 200

secondary visual cortex 107, 185
sensations 54, 60, 79, 86, 90, 109, 110, 121, 181
sensory leakage 127
Shadowlands 149
Shakespeare 120
shaman 35
Sheldrake, Rupert (Rupert Sheldrake, Dr. Sheldrake) 103, 104, 105, 106, 107, 108, 115, 116, 132, 152, 180, 194, 211, 220
Shermer, Michael (Dr. Shermer, Shermer) 130, 131, 132, 133, 134, 161, 214
Signs of Life 76, 163
single blind 70, 126
sitter 39, 41, 48, 55, 62, 65, 66, 68, 70, 84, 86, 96, 126, 128, 129
skeptic 73, 124, 127, 139, 164, 214
Skepticism 124
skeptics 40, 129, 134, 135, 145, 194, 207
skydiving 10
Smith, Gordon 75, 76, 189, 191, 206
smoke 66
Social Sciences 167, 185
social workers 119
Socrates 95, 184
sonar 89
Sony Walkman 30
sorority girls 160
soul ix, 62, 131, 132, 146, 148, 173, 184
sounds 40, 48, 55, 57, 58, 64, 65, 117, 121, 134, 162, 177
South America 38

space 141, 176, 177, 196
special chair 73, 74
Spirit Messenger 75, 76, 206
spirits 35, 90, 145, 146, 149, 151, 179, 191, 195, 203
Spirits --- They Are Present 86, 195
Spiritual 57, 59
spiritual hierarchy 144
standards of performance 91
Stanford 11, 116
stars 2, 44, 48, 143, 177
Stephen Hawking 36
Steve Grenard 53
Steve Jobs 117
Stevenson, MD, Ian (Dr. Stevenson, Dr. Ian Stevenson) 99, 100, 214
St. Francis 43
St. Francis of placeCityAssisi 43
subatomic particle 141
subjective 86, 157
suctioning device 11
sugars 138
Summerland 142, 144
survival of bodily death. 78
survival-of-consciousness 4, 124, 135, 152
survival of consciousness (SOC) 113
Susy 42, 55, 56, 63, 68, 69, 85
Susy Smith 41

T

Taiwan 1
Talking to the Dead, 132
Tears in Heaven 46
telepathic 98, 144, 151, 194
telepathically 147, 149, 151, 158, 161
telepathy 85, 144

Telepathy 152
temporal cortex 107, 108
terminal dendrite endings 110
TestingMediums (TMs) 53
The Afterlife Experiments 38, 175, 193, 211
The Afterlife Unveiled 145, 153, 200, 212, 220
The Airmen Who Would Not Die 209
The Electrical Evocation of Memories 107
The Fun of Dying 139, 198, 199, 212, 220
The Holographic Universe 36
The Matter Myth 36
theologians 170
The Oxford University Press Handbook of Psychology and Spirituality 110
The Presence of the Past 103, 115, 194, 211, 220
therapeutic 32
The Truth about Medium 51, 70, 195, 211
The Windbridge Institute (Windbridge Institute, WBI) 54, 90, 134, 213
Thorazine 22
throat 57, 64, 65, 96, 128
time 141, 144, 150
To Heaven and Back 168, 200, 201
Toy Story 43, 44, 48
traditional religions 137, 170
transition 118, 156, 195, 198
triple blind experiment 70
Trips 118
trite 167
Truncus Arteriosus 11, 186
tuberculosis 7, 186

Tucson 74
turkey baster 11
TV 38, 41, 105, 106, 161, 190
Twilight Zone 21

U

UC Santa Barbara 25
uncorroborated sitter ratings 129
unique 41
universal energy field 78
Universal Life Force 132
University of Arizona 38, 49, 50, 125, 195, 213
University of Colorado 25
University of Oregon 125
University of Virginia 99, 100
Uruguay 38
US Navy 1
Utah 38

V

Valium 22
venial sin 166
Veritas Research Program 50, 213
video playback 163
Virgil 89
visible-light spectrum 144
vision 32, 33, 35, 83, 117, 119, 120, 145, 150, 158, 175, 185
Visions 83, 117, 198, 211, 220
Visions, Trips and Crowded Rooms 117, 198

W

wannabes 90
warm reading 133
Washington Post 117, 215
waves of energy 141

Wayne 7, 8, 9, 38
Weirton 7
Weiss, M.D., Brian L.(Brian Weiss) 101, 207, 208
Where God Lives 110, 207, 212
white lie 166
Wikipedia 27, 31, 60, 101, 123, 125, 130, 138
Wilder Penfield 107
wild 'trips' 133
William James 37, 53
William James Fellowship 53
Woody 43, 48

X

Xenoglossy 100, 186, 214
X-Mas 73

Y

Yale 38
Yeshua 144

About the Author

Bill Kaspari was living a normal life even though he had lost his first son, who died as an infant. He had an engineering and business background and accepted, without giving the subject a great deal of thought, the modern scientific view of life that we are "here today, gone tomorrow" with no real vision of where we might go.

When he lost a son for the second time, his world was completely turned upside down. He had gotten so close to his son John, who was killed at the age of twenty-two, that he desperately needed to find answers to the question of whether or not we survive the death of our physical body, and if we do, what follows that physical death.

He began a search in the only way he could understand and accept—following scientific methods of inquiry. However, he soon found that mainstream science virtually ignores the subject, and treats life as though it is a coincidental result of evolution following the "big bang" and that there is no existence following death. Motivated by his deep need to "know" and also by the "so what—we all die" attitude he occasionally encountered, he began his search.

After years of reading what he could find on the subject, he became involved in the study of mediums, people who apparently have an innate ability to communicate with the deceased. He discovered, using emerging new scientific methodologies, that there clearly appears to be an existence beyond what we call death.

The result of this new understanding has had a profound, transformative effect on him. He now views life as a fantastic learning experience with unlimited opportunities to grow in both an intellectual and spiritual way.

CPSIA information can be obtained at www.ICGtesting.com
Printed in the USA
BVOW071546170413

318425BV00001B/5/P